S0-CAH-658

# LEAF PROTEIN:
its agronomy,
preparation, quality and use

IBP HANDBOOK No. 20

# LEAF PROTEIN: its agronomy, preparation, quality and use

Edited by
N. W. PIRIE

INTERNATIONAL BIOLOGICAL PROGRAMME
7 MARYLEBONE ROAD, LONDON NW1

BLACKWELL SCIENTIFIC PUBLICATIONS
OXFORD AND EDINBURGH

ISBN 0 632 08350 6

First Published 1971

*Distributed in the U.S.A. by*
F. A. DAVIS COMPANY
1915 ARCH STREET
PHILADELPHIA, PENNSYLVANIA

*Printed in Great Britain by*
BURGESS AND SON (ABINGDON) LIMITED
ABINGDON, BERKS
*and bound at*
THE KEMP HALL BINDERY
OSNEY MEAD, OXFORD

# Contents

# Foreword

The International Biological Programme which runs for the decade 1964 to 1974 was set up to study the 'biological basis of productivity and human adaptability' on a world basis. One of the seven sections into which the Programme is divided is entitled Use and Management of Biological Resources and this in turn is divided into several themes of which one is concerned with the development of new biological resources. This, since the beginning of IBP, has been under the leadership of C-G Heden of Stockholm and N. W. Pirie of Rothamsted, being based on co-ordinated research projects from a number of participating countries.

The theme was defined (*IBP News* No. 2 February 1965) thus: 'Investigation should be carried out on those unused and little used plant and animal products which might be converted into nutritionally exploitable materials by bio-engineering techniques.' It has been developed and the research projects which compose it have been guided and co-ordinated at three international meetings. The first was in 1966 at Warsaw, when a wide range of novel protein resources was discussed. The second was in 1968 at Stockholm at a symposium sponsored jointly by IBP and the Wenner-Gren Foundation which resulted in a volume of proceedings (Evaluation of Novel Protein Resources, edited Bender *et al.*, published by Pergamon). The third, which was devoted specifically to leaf protein, was held in December 1970 at Coimbatore and had as one of its objects the completion of the present handbook.

The development and use of leaf protein, which is potentially the most abundant of the novel protein resources, has been actively pursued within IBP under the stimulus of N. W. Pirie. This book brings together and reviews methods currently in use for production, extraction and processing, against an historical background, and in relation to the composition and nutritional value of leaf protein and its presentation and acceptability as human food. It also serves as a source book of information for those who may wish to develop the extraction and utilisation of leaf protein on a large scale.

IBP projects in leaf protein have been concentrated in Commonwealth countries and it was particularly appropriate that the Coimbatore meeting should have been supported by a generous grant to IBP by the trustees of the Commonwealth Foundation. I should like to acknowledge this with much appreciation and also to congratulate Mr Pirie and his collaborators for completing this handbook for publication within a remarkably short time of the formative meeting.

E. B. WORTHINGTON

# Introduction

Berzelius coined the name protein and defined the chemical category in 1838 (Hartley, 1951). Many resemblances had, however, been noticed by Rouelle in 1773 between animal substances and the coagulum that separates when leaf extracts are heated. His paper is now often quoted but does not seem often to have been read; a translation of it is therefore printed as an Appendix to this volume. In spite of early recognition, the proteins in leaves got little attention during the years in which the foundations of our knowledge of proteins were being laid by studies on egg white, milk, blood, and various seeds, especially those of the legumes. The presence of protein in leaves was fully recognized by those concerned with animal nutrition, but very little attention was paid to ways in which it could be extracted. Winterstein (1901) extracted protein from dried, ground leaves with dilute alkali, but sustained work did not start until 20 years later, when Osborne & Wakeman (1920), Chibnall & Schryver (1921) and Kiesel *et al* (1934) extracted protein from the fresh leaves of several species. Since then there has been a steady increase in the amount of work on leaf proteins; most of it is incidental to the study of photosynthesis, virus infection, or the general metabolism of the leaf.

There has been no comparable increase in interest in the practical use of extracted leaf protein. In 1924 and 1925 K. Ereky, the Minister for Development in the Hungarian government, tried to get articles published in Britain about a method he proposed for processing green crops on a large scale. He was deliberately inexplicit about both the method and the merits of the products made, but a patent (1926) was soon issued. The machine consisted of knives set on the opposed faces of a pair of coaxial truncated cones, and greenstuff was to be introduced in a stream of water which could be recycled to prevent too great dilution of the extract. Ereky did not describe how he separated the fibrous residue from the protein-containing extract, and he had some confused and largely erroneous ideas about the changes that took place in forage during autolysis and drying. He was, however, clear about the objective: to make from a fodder suitable only for ruminants a protein concentrate suitable for nonruminants. The advantages of thus partially bypassing the ruminant were stressed by Slade (1937), and he and Birkinshaw (1939) obtained a patent on a process by which leaf protein was precipitated from weakly acidic solution, although the method had been used by Winterstein and most other workers with leaf extracts. Slade & Birkinshaw described extractions in the laboratory only, but Goodall (1936) took out a patent for the use of full-sized sugar-cane rolls for extracting juice from leaves so that it could be evaporated and used as a vitamin preparation.

By 1939, therefore, the position had been fully defined; the quantity and extractability of protein in a few leaves was known, the advantages of extracting it were recognized, some methods for doing this were known, and so were the bulk properties of the protein. All that remained was to find out whether the production of leaf protein in bulk was practicable and whether the protein would prove to be as useful as theory suggested.

Fresh leaves, rubbed or comminuted so as to liberate most of the protein from the cells, yield a pulp that is easily handled in the laboratory but that is intractable in bulk. The conventional engineering approach to material like this is either to dry it so that it can be moved through the equipment on a current of air or to add water to it until the slurry will flow. Drying is expensive, and much of the protein in leaves denatures when the leaf is dried; this remains unextractable. To get a slurry that will flow satisfactorily, the water content of the pulp must be increased to 95% or more. If the crop contains 85% water when harvested, this necessitates adding to the crop four or five times its wet weight of water. Adding water (or recirculated leaf extract) on that scale is troublesome and complicates the later processes of coagulation, and the separation of the coagulum from the liquor. Therefore, although it is usual to extract leaf protein in the laboratory by mincing or grinding leaves with added water, this approach seems unsuited to large-scale work. It has, however, been advocated (Chayen *et al*, 1961).

The domestic meat mincer does an admirable job in the laboratory and with soft leaves; 375 W (half-horse-power) machines 8 cm in diameter will make 10–20 kg of pulp an hour. Larger mincers of the same basic pattern are unusable because they pack the charge too tightly and generate too much heat. Many types of screw expellers, designed for pressing oil out of fish or seeds, were tested and found to be unsatisfactory for the same reason. They work on a compacted mass and, with material as rough and unlubricated as most leafy crops, friction becomes excessive. Expellers are used successfully on citrus waste and they have been used on leaves (Casselman *et al*, 1965), but with the object of removing part of the water rather than extracting protein.

Because of the possibility that leaf protein could be used as a human food if war had led to blockade, various large-scale disintegrating devices were tested in 1940. Ball mills, rod mills, edge- and end-runner mills, and dough breakers were tried and found usable but unsatisfactory. Sugar-cane rolls, in spite of their use by Goodall, were also judged unsatisfactory because of heavy power consumption and the incomplete extraction achieved. They have, however, been reintroduced (Kohler & Bickoff, p. 69). Stamping mills of the type used to crush ores were not tried. The method may deserve investigation because laboratory work (Pirie, 1953) shows that pulping by impact is economical. It was with this background that the pulpers described here (p. 54) were designed. A proper comparison between their efficiency and that of sugar-cane rolls would be interesting.

The advantages to be expected from leaf protein production are fully dealt with in the section on agronomy in this volume. The advocates of leaf protein production do not claim that, in all circumstances, this method of producing protein is to be preferred to increased production of legume seeds or the newer protein-rich cereal seeds. It is a complement rather than an alternative to these methods of getting more protein. But it has advantages in regions where frequent rain makes it difficult to ripen a seed crop. Furthermore, when forage is being produced as an animal feeding stuff, it is unprofitable to apply the amount of fertilizer that would yield the maximum amount of crop because the crop would then contain more protein than a ruminant animal needs. If the excess protein were extracted for use by non-ruminants, and the residue were used as ruminant feed, optimal application of fertilizer would be justified.

The agronomic section also deals with the factors governing the selection of crops for extraction. There is no advantage in extracting protein from those leaves that are already used as vegetables. Few communities eat vegetables to an extent approaching the limit set by the human digestion; it would be preferable to make the eating of vegetables more popular. Trees are a possible source of leaves (Pirie, 1968b) in circumstances where collection is feasible; in most circumstances it would be impractical to collect them on an adequate scale. All the protein so far used in human or animal feeding experiments has been made from conventional crop plants and this state of affairs is likely to continue for several more years. The species and varieties used were, however, selected for reasons entirely unconnected with the extractability of the protein in their leaves. It is therefore reasonable to assume that other varieties, perhaps those at present rejected by plant breeders because of poor production of some such end product as a seed, will yield even better. There are also potentialities in wild plants. Not the unmanured rubbish of roadsides, but the same species given reasonable levels of husbandry. Toxicity is probably not an obstacle to their use because nearly all known toxic components of leaves would be extracted from the protein during the process of purification.

When the IBP was established in 1964, work on leaf protein was included in the program of the Use and Management (UM) section. The UK IBP committee arranged for a grant to design and make laboratory-scale equipment that could be used internationally to measure the extractability of the protein in local leaves. The grant also covered training in the use of the equipment. Leaf protein is now being investigated under IBP auspices in Ceylon, India, New Zealand, Nigeria, Pakistan, Sweden and UK. Work not connected with the IBP proceeds in several other countries.

Work with a bearing on the production and use of leaf protein is not restricted to the UM section of the IBP. In some countries it is part of the program of the Production Processes section and aspects of it will later be of interest to the Freshwater Productivity section. The procedures that should be adopted for gaining acceptance for leaf protein as a human food would, if they form part of the IBP program, be the responsibility of the Human Adaptability (HA) section. With this and other possibilities in mind, a joint HA–UM subcommittee was set up in the UK to consider the use of all the various novel foodstuffs that are now being investigated or proposed.

# Abbreviations

This list gives abbreviations other than the conventional ones for single elements, quantities, organizations, etc.

| | |
|---|---|
| ALV | available lysine value |
| BV | biological value |
| df | degrees of freedom |
| DM | dry matter |
| %E | percentage of the N in the leaf that appears as extracted protein |
| FM | fish meal |
| FYM | farm yard manure |
| GPV | gross protein value |
| HP | horse power |
| LP or LPC | leaf protein, or leaf protein concentrate. No distinction is intended between the two styles |
| NPK | nitrogen, phosphorus and potassium fertilizer |
| NPN | non-protein nitrogen |
| NPU | net protein utilization |
| NV | nutritive value |
| PER | protein efficiency ratio |
| PN | protein nitrogen |
| PVC | polyvinylchloride |
| SEM | standard error of the mean |
| SP | superphosphate |
| TCA | trichloroacetic acid |
| TD | true digestibility |

# Section I
# Agronomy

# 1

# Agronomic Aspects of Leaf Protein Production in Great Britain

D. B. ARKCOLL

## Introduction

It is necessary to know how much protein can be produced and extracted per unit of land in order to establish the full potential of the process. Consequently the agronomic as well as the processing variables that affect protein yield have been studied. The knowledge thus gained has been used to improve the yields of extracted protein from 400 kg/ha in 1956 to 2000 kg/ha in 1968.

**Selection of species**

Most crops grown in conventional agriculture have been selected for high dry matter production. Although leaf protein consists largely of the enzymes that synthesize dry matter, the ratio of protein to dry matter is not constant because wide differences in the activities of the metabolic systems of different species produce large variations in composition. Thus high dry matter yields can be produced by crops like maize and sugar-cane in spite of their low protein content. Fast fresh weight production is a more useful indicator because rapid cell division and extension usually implies a high protein content. Furthermore, rapid growth ensures a succulent plant comparatively free from secondary thickening and lignification, so that cells may be readily ruptured and the protein released mechanically.

Total nitrogen yield is a better guide to protein levels but may be misleading in some crops; for instance kale can have up to half its nitrogen in a non-protein form. With early machinery, the extraction from different species varied enormously and for many years (Byers, 1961; Rothamsted Experiment Station Annual Reports from 1952) this ease of rupture was in effect the main criterion for selection. It is now possible to extract 55–75% of the crop nitrogen as protein from most species unless they are very old and woody; selection can now depend on the rate of protein production. It is at present impractical to use a few fast-growing species because their sap is mucilaginous; this impedes the separation of juice from fibre (e.g. comfrey, bracken). A few others have acidic saps (e.g. dock, rhubarb) or contain so much tannin (e.g. *Desmodium uncinatum,* strawberry) that much protein precipitates in the pulp and is not extracted into the juice. Some precipitation occurs with most species probably due to the natural instability of chloroplastic proteins. This causes a considerable diminution in extraction if there is any delay between pulping and pressing.

A large number of species that grow rapidly have been examined with the aid of the IBP extraction unit (Davys & Pirie, 1969; Davys *et al*, 1969) to see if their protein can be extracted satisfactorily. These were either collected wild or were grown experimentally on 2 $m^2$ plots and are usually sampled when vegetative growth begins to slow down and flowers develop. Experience has shown that this is close to the age of maximum yield.

**Age at harvest**

Those species that grow especially fast and extract well are then grown in 200 $m^2$ plots which allows about 5 cuts to be taken at different ages for extraction with large-scale machinery. This gives the rate at which extractable protein is produced, the maximum yield and the regrowth potential after the early cuts. Non-leguminous crops with 8 weeks or less vegetative growth are usually top dressed with 628 kg/ha of Nitrochalk (21% N) and those with longer growth get 1256 kg/ha. All plots are given a liberal basal dressing of P and K (at least 110 and 160 kg/ha/annum) so that neither is limiting and crops are sown at very heavy seed rates (2–4 times normal agricultural rates) to ensure rapid ground cover so that early cuts can be taken to study the potential of regrowths.

The change in protein yield with age can be seen for a few crops in Fig. 1. Most protein is obtained by harvesting just before vegetative growth ends and floral growth starts. A few species continue to grow vegetatively after this stage (e.g. fodder radish, fat hen) in which case the optimum age is extended over a longer period or is slowly reached just before the plant withers (e.g. maize). From the data collected during the last 4 years it would seem that species unable to produce extractable protein at an average rate of about 8 kg/ha/day from the time of sowing to the optimum time of harvest are of little use in Britain. We have a growing season of about 200 days so that, if this rate is maintained, yields of at least 1600 kg/ha can be predicted.

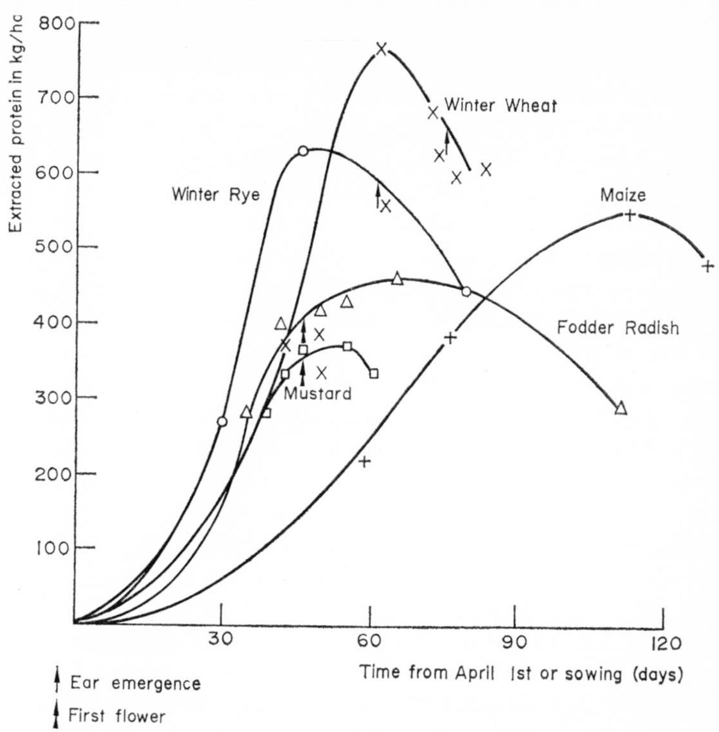


Figure 1.1. The effect of age on the protein yield of different crops.

The sudden fall in yield after the optimum age makes harvest timing extremely critical for many crops; a week early or late may make a difference of nearly 200 kg/ha in the final yield from wheat and rye. This rapid fall is caused by a sudden cessation of protein synthesis whilst proteolysis continues, and is exaggerated by the decline in extraction rate with age; leaves dry out and die as flowers develop and this lowers the extraction rate still further. Many factors influence this change in extraction rate and all are affected by climate, fertilizer and species. They include the change in PN to NPN ratio, the decline in juice to fibre ratio, and the change in character of the fibre which leads to less effective pulping of cells with thickened walls, and to greater absorption and filtration of juice on pressing.

Attempts to delay the quick fall in yield have been made by preventing the onset of flowering. Some success has been achieved by growing cereals and mustard at unusual times of the year and in unfamiliar latitudes, so that the day length and temperature are not those that initiate floral development. Unfortunately the growth is often less vigorous but the plants extract and regrow well. Use might be made of growth regulators that delay flowering and kinetins that promote protein synthesis and delay senescence. However, an easier solution is to use those species that regrow well.

**Regrowth ability**

Large yields are obtained from species that regrow well because the delay in floral development ensures good extraction from succulent young growth over a long period. Few cereals regrow vigorously but the maximum yield of wheat, barley and rye can be increased by harvesting when they are 30 cm tall and using the regrowth. Unfortunately the final yield takes longer to achieve and the rate of protein production is lowered. Lucerne and red clover have much more vigorous regrowth and usually yield about 1000 kg/ha when cut three or four times a year. More frequent cutting is not justified, unless the lower leaves start to die, because good extraction rates are achieved in these species up to the time of budding. Cutting too young can lead to a very wet pulp that is difficult to press.

Grasses are tough and were difficult to extract with early types of machine; this led to their rejection as suitable species. However, recent knowledge on how extraction rates are affected by age, nitrogen fertilizer, moisture content, second extractions and the speed and design of the pulper, has led to their reintroduction. Their ability to regrow is enormous and the high yield of 1670 kg/ha has been achieved in a dry year with cocksfoot cut five times and given 105 kg N/ha/cut (Fig. 2). The number and timing of cuts will depend on a compromise between the diminishing extraction rate, uptake of N and synthesis into protein, and the maintenance of an optimum leaf area to make maximum use of sunlight without starving the lower leaves and causing their premature senescence.

**Soil fertility**

All the nutrients known to affect a plant's growth may influence its protein content and some of them also alter the extraction rate by their effect on the juice to fibre ratio. A considerable lowering of protein yields has been shown on P, K and Mg deficient plots at Rothamsted. P and K are consequently applied liberally as a basal dressing at the start of the growing season. Magnesium is just beginning to become a problem on heavily cropped land

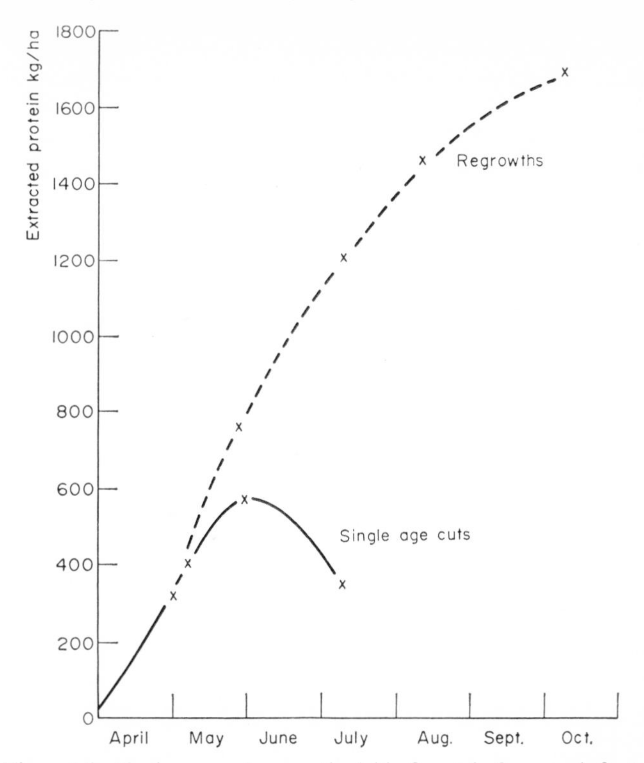


**Figure 1.2. Single age and regrowth yield of protein from cocksfoot.**

in Britain and it may soon be necessary to apply this as a basal fertilizer. Protein production will also make heavy demands on sulphur so that this and other trace elements may be needed on some soils.

The dominating element is nitrogen for it promotes abundant succulent growth. Its effect on those species with the highest rates of protein production are worth further study. These are usually grown at four or more levels of nitrogen to establish the optimum dressing. Protein yields respond to heavy dressing of this element; the extent can be seen for wheat in Fig. 3. The dry matter yield is not improved above 130 kg N/ha whereas protein yield increases up to 264 kg N/ha. Nitrogen not only increases the protein content of plants but usually improves the extraction rate by promoting luxurious growth. This latter effect is important with tough grasses so that the evaluation of unfertilized species growing wild can sometimes give misleading results.

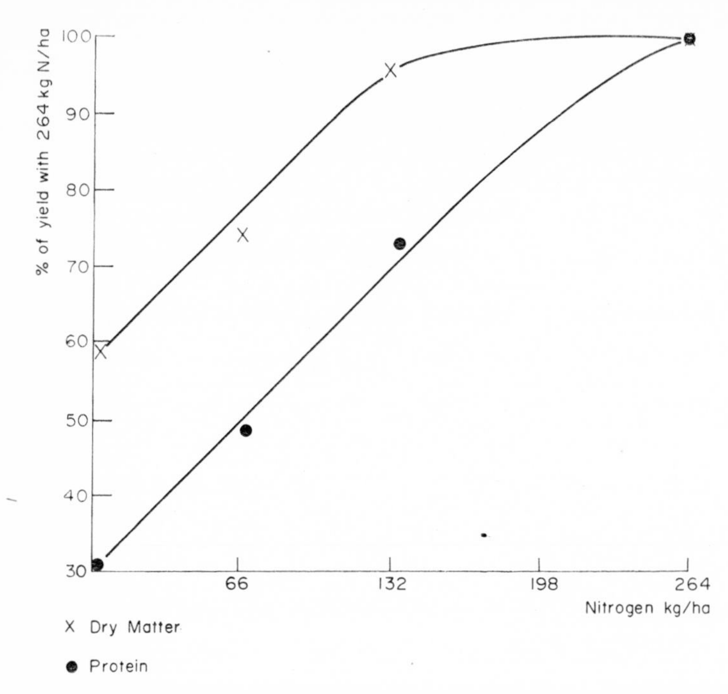


**Figure 1.3.** **The response of dry matter and protein yield of wheat to nitrogen at ear emergence.**

Finding the correct time to apply N is just as important as finding the optimum dressing. The interaction with climate is well known, for N is quickly lost by leaching if applied before plants are large enough to absorb it rapidly, while late application can result in an early period of N starvation from which plants may never recover. Our usual practise is to topdress with Nitrochalk just as plants are about to elongate rapidly. Those that germinate and establish fast can be fertilized at sowing. Uptake of nitrogen can be hastened by sowing thickly. Fodder radish has yielded 25% more protein when grown in 12·5 cm instead of 25 cm rows.

It is common practise to split the N topdressing for plants which have a long period of growth; for example, grasses are fertilized immediately after each cut. However, experiments on wheat showed no difference between a single dressing of 264 kg N/ha and two of 132 kg N/ha each. Split dressings may be justified when single very high levels damage seedlings and in those parts of the tropics where very high rainfall washes N away rapidly.

Nitrogen may be applied in many forms and experiments are constantly in progress to compare them. Ammonium ions are more readily synthesized into protein because the reduction from nitrate is avoided; this effect has been shown to increase protein levels in grass if an inhibitor is used to prevent nitrification within the soil (Nowakowski & Cunningham, 1966). Nitrochalk is a stable mixture of ammonium nitrate and limestone and helps to avoid the build up of acidity associated with ammonium sulphate. The protein content of plants can be increased by stimulating nitrate reductase with sub-lethal doses of the herbicide 'Simazine' but we have been unable to make practical use of this effect under good growing conditions.

Leguminous crops need no applied nitrogen. Lucerne and red clover can fix up to 300 kg N/ha in Britain and usually yield about 1000 kg of extracted protein/ha each year. Up to twice this amount has been fixed by clovers in warmer parts of New Zealand, and Singh (1967) has shown that irrigated lucerne is capable of yielding 3000 kg protein/ha in India. Although the protein yields of crops that respond well to nitrogen may be double that of the legumes it must be remembered that financially the cost of fertilizing to some extent offsets the gain.

**Use of the growing season**

The aim of this work is to increase annual protein yield by making the best use of the growing season. In Britain this only lasts for 200 days as very little growth takes place in the colder months. In many areas the season is restricted further by mild summer droughts in about 4 out of 5 years. An ideal crop would extend the growth period into the colder months yet be able to resist these droughts and efficiently utilize the greater radiation of the summer.

The large effect of climate on protein yield can be seen in Fig. 4. A warm spring allowed the 1968 crop to pick up N rapidly and synthesize it into protein before much secondary thickening had developed and impeded extraction. Experiments examining the effect of irrigation have shown improved uptake of N, synthesis of protein and better extraction from the more succulent and turgid plants.

The table shows that both red clover and lucerne make good use of the growing season although the latter grows better in dry years. Many of our best crops do not regrow well and so an effort has been made to use them by growing them in sequence experiments. The best combination found so far

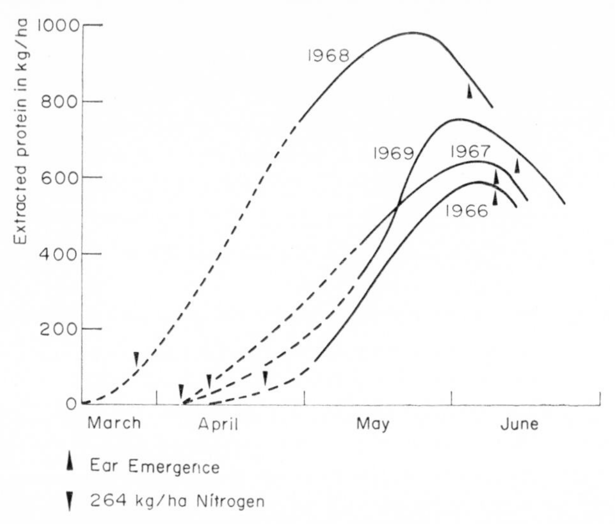


**Figure 1.4. The effect of season on the protein yield of wheat at different ages.**

is winter wheat, which yields well in spring, followed by two crops of either fodder radish or mustard which germinate fast and usually grow well in summer and autumn. Annual yields fluctuate between 1250 and 2000 kg/ha. Improvements in our knowledge of agronomic and processing variables now allow us to use grasses and the best of these, cocksfoot, gives extremely high annual yields as it grows early in spring and late into autumn. Such a crop is less vulnerable than sequence cropping to climatic variation (viz. 1969) and it should yield 2000 kg of extracted protein/ha in the wetter parts of the country when given 530 kg N/ha. Tropical grasses photosynthesize more efficiently, respond to higher levels of N, and grow all the year round. Hence they have a very high potential as is shown by napier grass which has produced 7200 kg of crude protein/ha/annum in Puerto Rico (Vincent-Chandler *et al*, 1959) so that extracted protein yields could reach about 5000 kg/ha.

A cropping scheme for the British climate might involve the extraction of excess protein and moisture from grass and lucerne before drying them. Foliar by-products like pea vines, potato haulms and sugar beet tops would fit in well in July, August and October and catch crops of mustard and fodder radish might be taken after early potatoes and peas.

TABLE

Total annual yields of leaf protein achieved at Rothamsted

| Year | Crop sequence and individual harvest yields of protein | Total annual protein yield (kg/ha) | Fertilizer (kg N/ha) |
|---|---|---|---|
| 1966 | Wheat 598; Fodder radish 332, 314 | 1244 | 510 |
| | Wheat 596; Mustard 298, 267, 98 | 1261 | 510 |
| | Wheat 598; Rape 250, 251 | 1099 | 510 |
| 1967 | Wheat+regrowth 433, 257; Fodder radish 391, 367 | 1448 | 528 |
| | Wheat+regrowth 433, 257; Mustard 295, 351, 99 | 1434 | 528 |
| | Wheat+regrowth 433, 257; Kale 654 | 1344 | 528 |
| 1968 | Wheat+regrowth 813, 285; Fodder radish 447, 461 | 2014 | 528 |
| | Wheat+regrowth 813, 285; Mustard 426, 490 | 2014 | 528 |
| | Wheat and vetch+regrowth 442, 372; Field bean leaves 469 | 1283 | 0 |
| 1969 | Wheat 765; Fodder radish 368 | 1133 | 395 |
| | Rye 625; Mustard 345, 216, 187 | 1373 | 528 |
| 1967 | Red clover+regrowths (irrigated) 334, 443, 301, 169 | 1247 | 0 |
| 1968 | Red clover+regrowths 447, 332, 155 | 934 | 0 |
| | Lucerne+regrowths 442, 407, 160 | 1009 | 0 |
| | Rye grass+regrowths 83, 271, 385, 53 | 792 | 0 |
| 1969 | Cocksfoot+regrowths 401, 362, 440, 253, 214 | 1670 | 528 |
| | Lucerne+regrowths 457, 383, 265 | 1106 | 0 |
| | Red clover+regrowths 233, 165, 127 | 525 | 0 |

## Conclusion

Much of the early work at Rothamsted was concerned with developing machinery to extract protein efficiently from crops specifically selected for ease of extraction. In the last few years improvements in the process have allowed the use of a wider range of species and preliminary attempts have been made to find the extent to which the more important agronomic factors affect extractable protein yields (Arkcoll & Festenstein, 1971). This knowledge has been used to maximize the annual yields of the best species selected so far. Yields have reached 1200 kg/protein/ha with legumes and 2000 kg/ha with crops given 530 kg/ha N in Britain. Respective yields of 3000 and 5000 kg/protein/ha seem possible in the wet tropics.

Species, age at harvest, N fertilizer, climate and seed rate are the most important factors governing protein yields. Improvements will be made in future by selecting better species, breeding them for better varieties and possibly by the use of growth regulators. The ideal crop will extend the growing season, respond well to N or be leguminous, and it will synthesize protein rapidly yet be slow to lignify and lay down secondary cell walls. Its regrowth will be vigorous and its foliage should resist disease yet contain little acid, tannin or mucilage. Although our best species are already in use in conventional agriculture there seems every reason to believe that wild species will be found with better characteristics for leaf protein production. This is even more likely in the tropics where a wider unused flora exists.

# 2

# The Yields of Leaf Protein that can be Extracted from Crops of Aurangabad

R. N. JOSHI

Work on leaf protein started in this department in July 1967; before 1969, a hand mincer was used to screen more than sixty species of wild and cultivated plants, thereafter the IBP pulper and press (Davys & Pirie, 1969; Davys *et al*, 1969) were used and only those experiments are quoted here. The pulper was driven at 3200 rev/min and juice was collected for 10 min.

**Agronomy**

*Medicago sativa* and *Vigna sinensis* were selected for intensive agronomic trials. These crops were planted at different dates on plots $10 \times 10$ m with three replications of three fertilizer treatments and were harvested at three ages. Seed rates were greater than normal and the plots were sown by hand in rows 0·31 m apart. Fertilizers were applied before sowing, 15 days later, and immediately after each cut. In connection with the production of leaf protein in quantity, the effects of (1) rhizobial inoculation of seeds, (2) top dressings of extra fertilizer N, (3) micronutrients and (4) simazine, on the yield of dry matter, crude protein and the extractability of the protein were studied. This report, however, includes only figures obtained on effects of rhizobial inoculations. The bacterial cultures were supplied by the Field Research Station, Bombay. The plots were arranged in randomized blocks with six replications and treated with only Farm Yard Manure (FYM) and superphosphate (SP).

Most of the other crops were grown on plots $5{\cdot}5 \times 4{\cdot}9$ m with row distance 0·31 m. Generous use of FYM was made in all these plots and details of variations in N and P treatments are given at the appropriate place.

All crops were harvested early in the morning and processed soon after.

Analyses and method of calculation were those of Byers & Sturrock (1965); the results stated as '% extraction' (%E) are the percentage of the total N

TABLE 1. Performance of various crops

| Latin name (1) | Date of sowing (2) | Date of harvest and age (3) | Leaf DM % (4) | Leaf N of DM % (4) | Protein N extracted % of total crop N (5) | Yield of protein extracted (kg/ha) (6) | Extractable protein (kg/ha/day) (7) | N % of Dm in fibre (8) |
|---|---|---|---|---|---|---|---|---|
| *Lablab niger* | 19 July 1970 | 24 August 1970 35 days | 14·0 | 4·0 | 48 | 151 }375 | 5·0 | 2·8 |
| | | 3 October 1970 regrowth 40 days | 15·0 | 4·04 | 42 | 224 } | | 2·8 |
| *Pennisetum purpureum*× *typhoideum* | 18 May 1969 | 13 July 1969 55 days | 15·5 | 2·07 | 30 | 87 }137 | 1·71 | 1·83 |
| (Gajraj) | | 7 August 1959 regrowth 25 days | 16·0 | 2·03 | 23 | 50 } | | 1·7 |
| *Sesbania sesban* Var. *picta* | 8 May 1969 | 23 June 1969 45 days | 17·2 | 4·0 | 56 | 192 }342 | 4·88 | 2·0 |
| | | 18 July 1969 1 regrowth 25 days | 15·0 | 4·17 | 59 | 150 } | | 2·0 |
| *Brassica nigra* | 17 October 1969 | 27 November 1969 40 days | 13·5 | 4·5 | 52·5 | 103 | 2·57 | 2·6 |
| *Dolichos uniflorus* | 22 June 1970 | 5 September 1970 74 days | 18·0 | 3·63 | 49 | 416 | 5·62 | 2·5 |

| | | | | | | | | | |
|---|---|---|---|---|---|---|---|---|---|
| *Amaranthus paniculatus* | 20 July 1970 | 5 September 1970 46 days | 16·4 | 3·71 | 59 | 208 | 583 | 6·47 | 1·68 |
| | | 19 October 1970 regrowth 44 days | 19·9 | 3·72 | 45 | 375 | | | 1·81 |
| *Atriplex hortensis* | 17 October 1969 | 4 December 1969 47 days | 13·5 | 4·5 | 72 | 173 | | 3·68 | 2·6 |
| *Coriandrum sativum* | 17 October 1969 | 4 December 1969 47 days | 12·0 | 3·8 | 43 | 78 | | 1·65 | 2·9 |
| *Rumex vesicarius* | 17 October 1969 | 9 December 1969 52 days | 8·6 | 5·3 | 43 | 105 | | 2·01 | 4·0 |
| *Spinacia oleracea* | 17 October 1969 | 9 December 1969 52 days | 9·2 | 4·3 | 53 | 109 | | 2·08 | 2·6 |
| *Trigonella foenum-graecum* | 21 October 1969 | 26 November 1969 35 days | 21·0 | 4·3 | 43 | 75 | | 2·14 | 3·23 |
| *Beta vulgaris* | 27 October 1969 | 21 December 1969 54 days | 11·0 | 3·0 | 52·5 | 87 | | 1·61 | 3·0 |
| *Brassica caulorapa* | 4 December 1969 | 24 January 1970 50 days | 15·1 | 3·3 | 57 | 84 | | 1·68 | 2·8 |
| *Brassica oleracea* var. *botrytis* | 4 December 1969 | 12 February 1970 69 days | 14·0 | 2·8 | 52 | 120 | | 1·73 | 2·2 |
| *Brassica oleracea* var. *capitata* | 4 December 1969 | 13 February 1970 70 days | 13·2 | 2·6 | 52 | 124 | | 1·77 | 2·2 |

TABLE 1—*continued*

| Latin name (1) | Date of sowing (2) | Date of harvest and age (3) | Leaf DM % (4) | Leaf N of DM % (4) | Protein N extracted % of total crop N (5) | Yield of protein extracted (kg/ha) (6) | Extractable protein (kg/ha/day) (7) | N % of Dm in fibre (8) |
|---|---|---|---|---|---|---|---|---|
| *Carthamus tinctorius* | 17 October 1969 | 26 November 1969 39 days | 10·0 | 5·0 | 57 | 84 | 2·15 | 2·9 |
| *Cicer arietinum* | 11 November 1969 | 17 December 1969 35 days | 20·0 | 4·4 | 50 | 69 | 1·97 | 2·0 |
| *Daucus carota* var. *sativa* | 17 October 1969 | 17 December 1969 60 days | 15·0 | 4·1 | 36 | 349 | 5·81 | 3·1 |
| *Raphanus sativus* | 17 October 1969 | 21 December 1969 64 days | 11·0 | 3·96 | 54 | 271 | 4·23 | 2·5 |
| *Lycopersicon lycopersicum* | 17 October 1969 | 29 January 1970 103 days | 13·0 | 2·9 | 54 | 182 | 1·76 | 3·06 |
| *Tithonia tagetiflora* | 28 May 1969 | 8 July 1969 40 days | 17·5 | 2·8 | 72 | 342 | 8·55 | 1·0 |

in the crop that appears as extracted, washed, protein N. The leaves tested were grouped in five categories.

(a) *Cover crops, fodders and green manures.* Earlier work at other centres (Byers, 1961; Byers & Sturrock, 1965; Singh, 1967; Stahmann, 1968; Samuel, personal communication) had indicated that these crops, most of them leguminous, are suitable. Four fodders and one hedge plant were found quite promising (Table 1). *Lablab niger* was planted at the end of the summer 1969 and the extractability of protein N and the yields per hectare were measured during the monsoon. The crop received 60 kg P/ha and yielded 220 kg protein when cut twice (45+30) in 75 days. The crop sown in July 1970, received 120 kg N/ha and yielded 375 kg protein in 75 days when cut twice. The %E varied between 35 and 49 during 1969 and between 42 and 48 during 1970. Gajraj (EBH–24) yielded 137 kg extractable protein in 80 days when cut twice (55+25); %E was 23–30. Among the grasses cultivated in this region as cattle feed this hybrid appears most promising and work on the effects of fertilizers and frequency of cutting on the protein yields has been started.

*Sesbania sesban* var. *picta* is an important cover crop and hedge plant of this region. The plant yielded 342 kg extractable protein/ha when cut twice (45+25 days), %E at the two cuts was 56·2 and 58·8 respectively. Some agronomic trials were started on this crop during 1969 and these continue.

Lucerne was sown in March 1969 which is an unusual period for its cultivation in this region and the effects of three fertilizer treatments and three different frequencies of cutting on total yield of protein in 180 days were studied. Plots treated with NPK received in all 125 kg N, 500 kg P and 125 kg K/ha, while those treated with FYM+SP received 4000 kg of FYM and 500 kg of P per ha. Lucerne was sown again in October 1969—the usual time for cultivation—and similar studies were carried out. Results presented in Table 2 are for plots which received NPK and which were cut six times within 180 days (age cut 50 days after sowing and five successive cuts each after 26 days). It is apparent from this table that more than 3 tons of protein could be extracted from lucerne per hectare in one year. FYM+SP plots yielded, on average, more than 2·7 tons. A crop cultivated at the usual time showed better growth and yielded about 500 kg more protein than the one cultivated in summer. The average extractability of total N and protein N was high during winter when it was possible to extract as much as 70% total N from the leaf, whereas poor extractabilities of protein N were observed during the hot days of summer. The whole trial is being repeated; during the

TABLE 2. Yields of dry matter and extracted protein from *Medicago sativa*

| Date of sowing | Type of cut | Yield of dry matter (kg/ha) in plots treated with NPK | Yields of protein extracted (kg/ha) in plots treated with NPK |
|---|---|---|---|
| (1) | (2) | (3) | (4) |
| 15 March 1969 | 2 Age | 1716 | 188 |
| | 1 regrowth | 2050 | 225 |
| | 2 regrowth | 1784 | 219 |
| | 3 regrowth | 2070 | 231 |
| | 4 regrowth | 2025 | 200 |
| | 5 regrowth | 2078 | 239 |
| | Total 180 days | 11723 | 1302 |
| 1 October 1969 | 2 Age | 1830 | 276 |
| | 1 regrowth | 2280 | 311 |
| | 2 regrowth | 1931 | 317 |
| | 3 regrowth | 2293 | 314 |
| | 4 regrowth | 2380 | 328 |
| | 5 regrowth | 2072 | 266 |
| | Total 180 days | 12786 | 1812 |
| 13 March 1970 | 2 Age | 1159 | 148 |
| | 1 regrowth | 1000 | 106 |
| | 2 regrowth | 1553 | 163 |
| | 3 regrowth | 1969 | 197 |
| | 4 regrowth | 2019 | 245 |
| | 5 regrowth | 1522 | 205 |
| | Total 180 days | 9222 | 1064 |

first half of 1970 the crop yielded 1064 kg protein with NPK and 916 kg protein with FYM+SP. Results on this crop are encouraging and we think that once the crop gets established, by keeping it for 1·5–2 years, with additional dressings of N and P and better control of pests after the monsoon, it should be possible to get yields between 3500–4000 kg/ha/year. Trials on the effects of rhizobial treatments have shown (Table 4) that it is possible to increase the protein content of lucerne by about 15%.

(b) *Leaves taken from crops which are cultivated for the production of seed rather than leaf*. This group included a few cash crops and preliminary screening with a hand mincer suggested that they are excellent sources of easily

TABLE 3. Yields of dry matter and extracted protein from *Vigna sinensis*

| Date of sowing | Type of cut | Yield of Dm (kg/ha) in plots treated with NPK | Yield of protein extracted (kg/ha) in plots treated with NPK |
|---|---|---|---|
| 7 May 1969 | 3 Age | 2285 | 294 |
| | 1 regrowth | 2910 | 349 |
| | Total 80 days | 5195 | 643 |
| 3 November 1969 | 3 Age | 1034 | 156 |
| | 1 regrowth | 841 | 114 |
| | Total 80 days | 1875 | 270 |
| 22 February 1970 | 3 Age | 1872 | 264 |
| | 1 regrowth | 2204 | 300 |
| | Total 80 days | 4076 | 564 |
| 25 May 1969 | 3 Age | 2637 | 356 |
| | 1 regrowth | 3981 | 491 |
| | Total 80 days | 6618 | 847 |

extractable and good quality protein. *Vigna sinensis* was chosen for intensive agronomic trials. The species was planted in three seasons of a year. During each season the NPK plots received 60 kg N, 240 kg P and 60 kg K/ha and the others 1850 kg FYM+240 kg P. The results presented in Table 3 are for NPK plots cut twice (45+35) within 80 days. During monsoon 643 kg, during winter only 270 kg and in summer 564 kg protein could be obtained. The plant did not show a good response to the application of FYM+SP. In 1970, the yield of extracted protein improved to 847 kg in 80 days during monsoon which represents more than 10 kg of extracted protein per ha/day. This plant, therefore, is best suited for cultivation with NPK during the rainy season. The average extractability of total N for all the three seasons varied between 65 and 77% when harvested at 45 days, and between 52 and 65 at the first regrowth 35 days later. Rhizobial inoculations of seed (Table 4) increased the protein yield by 12 to 19%.

Byers & Sturrock (1965) reported that mustard gives a quick return in extracted protein. We could, however, obtain only 103 kg protein in 40 days. It may be noted that this yield was obtained when the crop received only 10 kg N/ha. *Dolichos uniflorus* yielded 416 kg protein in 74 days at the rate

TABLE 4. Effect of rhizobial inoculation on crude protein and protein extracted from the leaves of *Medicago sativa* and *Vigna sinensis*

| Name of plant | Date of sowing | Yield of crude protein (kg/ha) | | Yield of protein extracted (kg/ha) | |
|---|---|---|---|---|---|
| | | Control | Inoculated | Control | Inoculated |
| *Medicago sativa* | 15 September 1969 | 1444 | 1719 | — | — |
| *Vigna sinensis* | 22 January 1969 | 611 | 715 | 367 | 432 |
| | 28 January 1970 | 719 | 851 | 373 | 403 |

of 5·62 kg/ha/day, whereas *Eleusine coracana* yielded only 70 kg during the same period.

(c) *Vegetables of the region. Atriplex hortensis, Spinacia oleracea, Rumex vesicarius, Trigonella foenum-graecum* and *Coriandrum sativum* yielded 173, 109, 105, 75 and 78 kg protein respectively. The extractable protein N was 72, 53·2, 43, 42·6 and 43·3% of the crop N respectively. *Amaranthus paniculatus* was specially grown for leaf protein extraction. The crop received 125 kg of fertilizer N/ha. The plant was somewhat mucilaginous but showed good extractability of protein N and yielded 583 kg protein when cut twice within 90 days which is 6·42 kg/ha/day.

(d) *Leaves classified as by-products.* This group consisted mainly of leaves taken from crops at the normal harvesting time or the leaves which are removed from certain crops at an early stage of their growth. In the former category, the leaves are abundant but mature and usually have high dry matter and low N content whereas in the latter, the leaves are tender, contain less dry matter but more N.

Groundnut (*Arachis hypogaea*) is the main oil yielding crop of this region. The leaves from two varieties, viz. TMV–2 and SB–II, were removed after the nuts were harvested and were very dry and withered; nevertheless, more than 27% N was extractable. 34 kg and 36 kg protein was extracted from these two varieties.

The by-product vegetation of four vegetable crops was found promising. The most interesting plant was tomato in which, %E exceeded 54 and the plant yielded 182 kg protein after the fruits were plucked three times. Leaves of carrot (*Daucus carota* var. *sativa*) harvested somewhat earlier, yielded

349 kg of protein; whereas those of radish (*Raphanus sativus*) and *Beta vulgaris* gave 271 and 87 kg protein. The %Es were 35·7, 53·7 and 52·5 respectively. Other vegetables and fruit plants, especially cucurbits giving lush foliage, are being investigated.

Singh (1970) has shown that nutritive values of leaf proteins from crucifers are even better than those of lucerne. Leaves of knol khol, cauliflower and cabbage were pulped after the harvest of the main crop and yielded 84, 120 and 124 kg protein/ha in 50, 69 and 70 days respectively.

*Cicer arietinum* and *Carthamus tinctorius* are generally cut once by the farmers to encourage more vigorous growth. The early cuts, which are normally used as cattle feed or sold as a vegetable, were examined for protein production. The former plant yielded 69 kg protein (%E=50·2) and the latter 84 kg protein (%E=57·4) in 35 and 39 days respectively.

It was impossible to express juice from the pulp of sweet potato leaves by our press although earlier screening had indicated a %E of 47. Some agronomic trials suggested that this crop needs heavy dressings of NPK. However, in plots which received 100 kg N/ha it yielded only 147 kg protein in 90 days. It must be mentioned here that only one extraction was carried out with pulp squeezed by hand. The cut reported here was taken before the harvest of the main product.

(e) *Miscellaneous*. This group includes twenty-eight weeds processed through the pulper and press. They are abundant during the kharif (rainy) and rabi (winter) seasons. In most of these it was not possible to measure the yields but an idea about their extractability and yield per ton of fresh vegetation could be obtained. Among wild plants *Cassia occidentalis, Cassia obtusifolia* and *Crotalaria willdenoviana*, which are abundant during the rains, were found suitable. A most interesting plant is *Parthenium hysterophorus* which has become a nuisance all over western Maharashtra. At the flowering stage %E exceeded 49; it yielded 14 kg protein per ton of fresh vegetation and this protein contained 9% N. Another sunflower-like plant often cultivated for its beautiful flowers—*Tithonia tagetiflora*—appears very promising (Table 1). The extractability of total N from this plant was as high as 83·9%; N content of protein being 9·7% and the yield of protein 342 kg/ha in 40 days. Other weeds such as *Digera* sp., *Launaea* sp., *Achyranthes* sp., *Peristrophe* sp. and *Amaranthus* sp. were found good and results on these are published elsewhere.

## Summary

The results presented indicate that vegetation from Marathwada offers an encouraging source for leaf protein production. They also show that weeds and wild plants, most of which go to waste, by-product leaves, and certain forage crops can be profitably employed. The results suggest that by using a suitable succession of crops (in which plants such as cow pea, *Amaranthus* sp. or *Tithonia* are cultivated) or by growing a perennial fodder like lucerne alone, it should be possible to get 3500 kg protein per hectare per year. Further data for 1- or 2-year periods would put this investigation on a sound basis for recommending plants for large-scale production.

# 3

# A Survey of Other Experiments on Protein Production

N. W. Pirie

No experiments covering such long periods or so many species have been published from sites other than the UK and Aurangabad. Many other experiments were, however, presented and discussed at the meeting in Coimbatore. Much of this material is, as yet, unpublished; when referred to here, this point is noted.

In Mysore, Singh (unpublished) used a batch extractor (Davys & Pirie, 1963) because he is primarily interested in making batches of protein large enough for feeding experiments. This unit is known to be unsatisfactory because the unavoidably long contact between the extracted protein and the pulped material causes serious loss of protein. The yields are, therefore, only about 60% of what would have been attained in more efficient extraction equipment (Davys & Pirie, 1969). The results are, however, worth recording, because the batch extractor is still being used for producing protein in India, New Guinea and Nigeria.

The yields from lucerne from different plots at different times during a period of over 4½ years are given in Table 1. The regrowths were harvested several times, as mentioned therein. The wide variation in yield between the first two plots (A and B) and others (C, D, E and F) were due to differences in the cultural, agronomic and supervisory background of the plots used. Plots A and B were on fertile land under cultivation for many years and plots C, D, E and F were newly brought under cultivation having earlier been under scrub vegetation. The former two plots had adequate irrigation and agronomic attention under experienced farmers, while the latter were inadequately tended. However, it was obvious that even under sub-normal cultivation, considerable amounts of extracted edible leaf protein could be obtained from lucerne. The fibrous residue would be suitable for direct use as bulk fodder for ruminants. In contrast to established practice in countries with a developed dairy industry, where lucerne persists as a satisfactory

TABLE 1. Batch production of leaf protein from lucerne (*Medicago sativa*)

| Plot and number of harvests of regrowth | Period* (days) | Yields of vegetation, batch processed materials | | | | | Computed per hectare yields of batch extracted food protein (kg/ha) |
|---|---|---|---|---|---|---|---|
| | | Area ($m^2$) | Fresh vegetation (kg) | Leaf protein DM (kg) | Food protein † (kg) | Fibrous residue dry (kg) | |
| Plot A 11 harvests between 9.11.1964 and 3.5.1965 | 196 | 133 | 1300 | 16 | 9 | 110 | 675 |
| Plot B 10 harvests between 11.11.1964 and 26.4.1965 | 187 | 133 | 1340 | 16 | 9 | 125 | 675 |
| Plot C 18 harvests between 6.12.1966 and 4.12.1967 | 365 | 480 | 2900 | 38 | 21 | 260 | 420 |
| Plot D Sown on 5.1.1967; 17 harvests between 13.2.1967 and 26.12.1967 | 315 | 500 | 3650 | 46 | 26 | 330 | 520 |
| Plot D Continued 18 harvests up to 3.2.1969 | 404 | 500 | 2250 | 33 | 18 | 205 | 360 |
| Plot E Sown on 10.3.1967; 7 harvests between 24.4.1967 and 29.8.1967 | 127 | 500 | 1700 | 24 | 13 | 155 | 260 |
| Plot F Sown on 2.7.1968; 14 harvests between 19.8.1968 and 26.5.1969 | 279 | 490 | 2400 | 35 | 20 | 215 | 400 |

* Between dates of first and last harvest.
† Food protein = N in the crude protein × 6.0.

perennial legume fodder crop for several years, experience in India shows a decline in yield from the second year onwards. This was also shown in the results presented here; the total vegetation-regrowths and the extracted leaf protein decreased on continuing the harvests into the second year (Plot D). Lucerne has therefore no special advantage over other annual crops in India. No explanation of this difference has been suggested. Research should be undertaken to find out whether this early failure of lucerne is a consequence of the varieties used, the climate, nutrient deficiency, virus or other infections, or infestation by nematodes and other soil predators.

The effects of irrigation and differences in the frequency of cutting were studied in New Zealand with two varieties of lucerne (Vartha & Jones, unpublished). During the period 8/9/69 to 7/5/70 the irrigation treatment comprised seven applications of 15·4 cm between 18 November and 19 February. Rainfall from September 1969 to May 1970 was 41·5 cm compared with the 30-year average of 47 cm. For spring and summer, monthly rainfall was average only in December and January.

Wairau lucerne, with 5-week intervals between harvests in the period 8/9/69 to 7/5/70, gave the following yields of herbage dry matter, protein extracted by the IBP unit (Davys & Pirie, 1969) and precipitable by TCA, and residue:

| | Herbage | Protein | | Residue | |
|---|---|---|---|---|---|
| | (kg DM/ha) | kg DM/ha | % of herbage | kg DM/ha | % of herbage |
| Irrigated | 19768 | 2341 | (11·8) | 13624 | (68·9) |
| Dryland | 14322 | 1504 | (10·5) | 9881 | (69·0) |

The ranges in %E and the % of the crop N that was NPN were:

| | % N | % NPN |
|---|---|---|
| Irrigated | 38·8–57·9 | 6·7–11·2 |
| Dryland | 41·7–60·1 | 6·0– 9·5 |

Effects of summer irrigation on herbage, extracted protein, and residue yields with 5-week intervals between harvests, expressed as % increase from irrigation, were:

| | 16 December | 20 January | 25 February |
|---|---|---|---|
| Herbage | 59 | 41 | 33 |
| Extracted protein | 67 | 51 | 45 |
| Residue | 58 | 40 | 28 |

Extractable total N % was also increased by irrigation:

| | 16 December | 20 January | 25 February |
|---|---|---|---|
| Irrigated | 56·5 | 51·8 | 57·9 |
| Dryland | 48·6 | 45·4 | 51·2 |

Total N % of the leaf and stem at 25 February was increased by irrigation.

Comparisons between lucerne varieties for herbage yield from 20/1/70 to 14/5/70 showed that Wairau outyielded Rhizoma lucerne, especially with infrequent cutting.

Yields in kg DM/ha were:

| | Wairau | | | Rhizoma | | |
|---|---|---|---|---|---|---|
| | 3 wk | 5 wk | 7 wk | 3 wk | 5 wk | 7 wk |
| Irrigated | 3867 | 6749 | 6868 | 3663 | 4570 | 4958 |
| Dryland | 3318 | 5515 | 5889 | 3139 | 3831 | 3888 |
| % difference | 14 | 18 | 14 | 14 | 16 | 22 |

For both lucerne varieties, irrigation had similar effects on herbage yields under all cutting treatments.

Extracted protein and residue yields for Wairau lucerne only, for the period from 20/1/70 to 14/5/70, were:

| | 3 wk | | 5 wk | | 7 wk | |
|---|---|---|---|---|---|---|
| | kg/ha | % of herbage DM | kg/ha | % of herbage DM | kg/ha | % of herbage DM |
| Extracted protein | | | | | | |
| Irrigated | 521 | 13·5 | 912 | 13·5 | 665 | 9·7 |
| Dryland | 414 | 12·5 | 715 | 13·0 | 646 | 11·0 |
| Residue | | | | | | |
| Irrigated | 2107 | 54 | 4354 | 65 | 4888 | 71 |
| Dryland | 1685 | 51 | 3576 | 65 | 3924 | 67 |

Total N % in leaf tended to be higher under irrigation.

The substantial improvement in herbage yield with irrigation was paralleled by the yield of extracted protein. Although leaf and stem N %, and % extractable N, were higher with irrigation in summer, for the total period of growth, yields of protein as percentages of herbage yields were similar under irrigation and dry land conditions.

Protein yield was at a maximum (912 kg in 115 days) at 5 weeks interval between harvests. Herbage yield was increased with longer intervals but leaf and stem N % and % extractable N were decreased. Attainable yields in U.S.A. are given by Oelshlegel *et al.* (1969a).

Rao, Deshpande & Perur (unpublished) extracted protein from field-grown *Lablab niger* with a domestic meat mincer in Bangalore. Phosphate (up to 66 kg/ha) increased the yield and N content after 105 days growth, and trace elements increased the yield a little more. The extractability of the protein was not affected. This work is a prelude to a more detailed study using the IBP extraction unit. This unit was used by Gunetileke (unpublished) in Ceylon to extract protein from the local 'spinach'—*Basella alba.* This leaf is esteemed as a green vegetable and it could be more extensively eaten; it can be argued therefore that there is no advantage in extracting protein from it. Gunetileke argued that protein made from less conventional leaves will be more readily accepted if people become familiar with the idea of using leaf protein through eating protein that carries a flavour to which they are accustomed. There were ten plants per square metre, they were still growing after 47 days but the extractability of the protein declined sharply after 30 days; at this point 7 g of protein was extracted from each plant. In Ceylon it should be possible to take ten harvests in a year, the annual yield would, on this basis, be 7 tons of dry extracted protein. This is the largest yield that has been claimed and it would be accompanied by fibrous residue and extracted NPN, both of which could be used in other ways (p. 135) to make more protein. Some other plants that are widely used as green-manures or cover-crops, and the leaves that are discarded when vegetables are harvested in the conventional manner, were extracted by Singh (unpublished), in the batch extractor that he used with lucerne. The results are set out in Table 2. All these leaves were from crops grown on unsystematically cultivated land within the Central Food Technological Institute (Mysore); the yields were therefore probably smaller than those that would be expected under good farming conditions. Mucilaginous vegetation, such as sannhemp (*Crotalaria juncea*) could not be used in the batch extractor.

Work at Rothamsted (p. 9) and Aurangabad (p. 19) shows that there is no reason to restrict attention to plant species for which there are accepted methods of large-scale cultivation. Besides conventional crops, several unconventional species were examined by Lexander *et al* (1970).

The plants used in these experiments were grown in a glasshouse with some supplementary artificial light. The grasses were harvested after 7 weeks and the other plants after 10–12 weeks. All were then frozen so that plants with a uniform background could be studied. The conclusions may not therefore apply to fresh material because the protein in some leaves is coagulated by freezing. There is no evidence that protein is ever made more soluble by

TABLE 2. Batch production of protein from by-product leaves

| Common names | Latin names | Period* (days) | Yields of vegetation, batch processed materials: Area ($m^2$) | Fresh vegetation (kg) | Leaf protein DM (kg) | Food protein† (kg) | Fibrous residue dry (kg) | Computed per hectare yields of batch extracted food protein (kg/ha) |
|---|---|---|---|---|---|---|---|---|
| *Green manuring crops* | | | | | | | | |
| 1. Pillipesera | (*Phaseolus trilobus*) | 58 | 348 | 220 | 4·7 | 1·7 | 24 | 49 |
| 2. Kolinji | (*Tephrosia purpurea*) | 57 | 644 | 126 | 4·5 | 1·9 | 12 | 30 |
| 3. Dhaincha | (*Sesbania cannabina*) | 42 | 322 | 380 | 4·5 | 2·4 | 29 | 75 |
| *Crucifer vegetable crops* | | | | | | | | |
| 4. Cauliflower | (*Brassica oleracea* var. *botrytis*) | 68 | 88 | 240 | 1·9 | 0·9 | 15 | 102 |
| 5. Knolkhol | (*Brassica oleracea* var. *caulorapa*) | 76 | 161 | 390 | 3·5 | 1·8 | 25 | 112 |
| 6. Turnip | (*Brassica rapa*) | 62 | 478 | 350 | 2·9 | 1·5 | 14 | 32 |
| 7. Radish | (*Raphanus sativus*) | 37 | 290 | 200 | 1·2 | 0·5 | 8 | 17 |
| 8. Cabbage | (*Brassica oleracea* var. *capitata*) | 104 | 94 | 176 | 1·6 | 0·5 | 16 | 53 |
| *Other vegetable crops* | | | | | | | | |
| 9. Potato | (*Solanum tuberosum*) | 100 | 370 | 190 | 3·2 | 1·4 | 10 | 38 |
| 10. Beetroot | (*Beta vulgaris*) | 73 | 300 | 510 | 4·7 | 1·8 | 27 | 60 |

* From sowing to harvest.

† Batch-extracted leaf protein dry matter (DM) N × 6·0 = Food protein.

freezing: the yields of extracted protein may therefore be regarded as minima. The frozen leaf was thoroughly disintegrated and the juice extracted with a hydraulic press. Juice expressed from heavily compacted pulp is to some extent depleted of chloroplasts and their fragments. This also will diminish protein yields.

Measurements of total and extractable protein N are given for twenty-four species, each tested at three levels of fertilizer N. The conclusions drawn are '. . . grasses (*Festuca pratensis, Dactylis glomerata, Lolium perenne*) and leguminous species (*Medicago sativa, Trifolium pratense, Vicia sativa*) had high protein nitrogen percentages even at low fertilizer levels, whereas nitrophilous species (*Chenopodium quinoa, Amaranthus caudatus, Urtica dioica*) obtained high protein nitrogen percentages by responding to heavy application of fertilizers. Some of the other species (*Melandrium rubrum* and *Cirsium oleraceum*) also had high protein nitrogen percentages even with low application of fertilizers.'

In the cultivation boxes, the distance between plants was adjusted for each species to what seemed most suitable for optimal use of the soil area without too much mutual shading. Thus, the values of *protein produced/m*$^2$ of soil may be used to compare the protein production capacities of the species when cultivated in a similar way. In Fig. 1, the complete columns represent grams of protein produced/m$^2$ at the highest fertilizer level. Only one species, *Melandrium rubrum,* produced more protein/area at the lower fertilizer levels (37 g v. 27 g at the highest fertilizer level). *Helianthus annuus* produced substantially more protein/area than any other species, followed by *Amaranthus caudatus* and *Atriplex hortensis*. The highest protein production per area was about ten times the lowest one. Some of the fertilizer-responsive species might have responded to still heavier fertilizer applications with even higher protein production, since they had not reached their maximum level of produced protein at the highest fertilizer level of this investigation . . . . The highest extractability was displayed by the Chenopodiaceae and among them, *Chenopodium urbicum* had the best value, i.e. 80%. Among the other species, only *Melandrium rubrum* and *Trifolium pratense* had as high extractability as the Chenopodiaceae. The lowest value, 5%, was found for *Cirsium oleraceum*. In several cases, the extractability was influenced by the fertilizer level, but the species differed in their type of extractability response to fertilizer.

The amount of extracted protein obtained/unit area at the highest fertilizer level from the different species and varieties can be compared using the solid

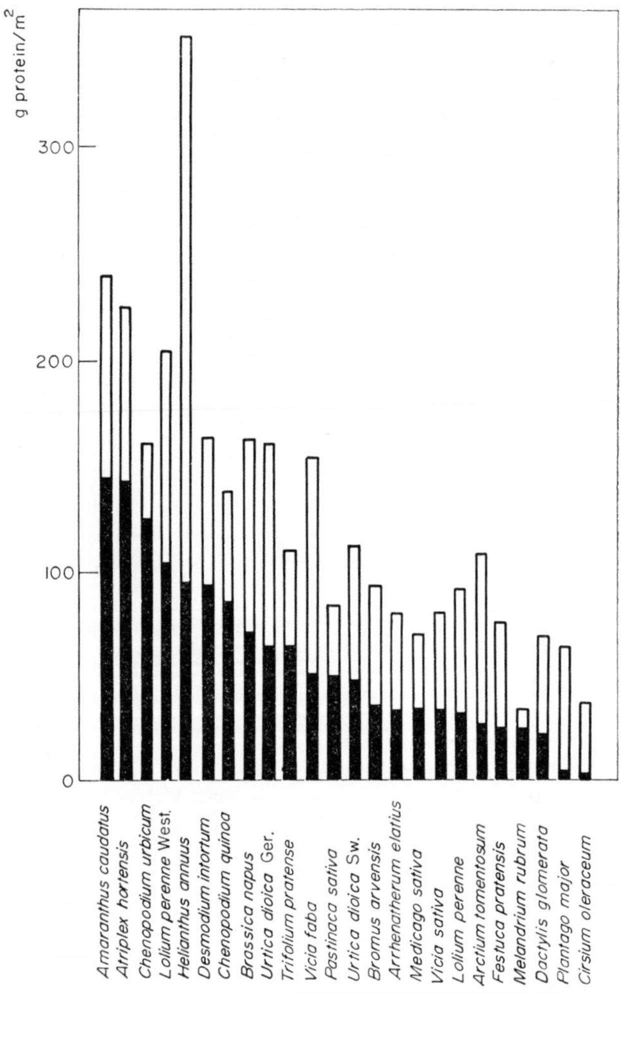


Figure 3.1. Species variation in g protein produced/m$^2$ (complete columns) and g protein extracted/m$^2$ (solid part of columns). All values are the means of two harvests.

parts of the columns of Fig. 1. The largest amounts of protein that could be utilized as protein concentrates were extracted from *Amaranthus caudatus Atriplex hortensis* and *Chenopodium urbicum*, which yielded between 125 and 143 g of extracted protein/m$^2$. Fig. 1 summarizes these results.

Using fresh material, and an IBP extraction unit (Davys & Pirie, 1969; Davys *et al*, 1969), Matai, Bagchi & Raychaudhuri (unpublished) undertook a somewhat similar study. Eleven species, *Hibiscus sabdariffa, Corchorus olitorius, C. capsularis, Pennisetum polystachyon, P. pedicellatum, Trillium* sp., *Mikania scandens, Alocasia indica, Helianthus annus, Pachyrhizus angulatus* and *Ipomoea batatas* were from cultivated plots in the Indian Statistical Institute, Calcutta. The rest of the species were collected from abandoned plots or lakes. Wild plants were mostly collected when mature as that eased identification.

The plants are divided into six groups: (A) water weeds, (B) plants growing in marshy land, (C) plants giving slimy pulp, (D) grasses, (E) wild plants, and (F) cultivated plants. Some details about the more interesting species are given in Table 3.

(A) *Water weeds. Eichhornia crassipes* (water hyacinth), *Pistia stratiotes* (water lettuce) are not used; *Ipomoea reptans* is used as a vegetable locally on a small scale. These plants do not extract well but they should be exploited for leaf protein production because (a) they grow throughout the year (except in the hot summer months May–July), (b) much effort and money is devoted to their removal, and (c) they are abundant, 50,000 kg/ha, 20,000 kg/ha and 25,000 kg/ha respectively as standing crop. Annually, the yield may well be 120,000 kg/ha, 70,000 kg/ha and 80,000 kg/ha. *Pistia* has the disadvantages that it (a) causes skin irritation during pulping and (b) the N percentage on crude protein is small.

(B) *Plants growing in marshy land. Ipomoea caruii* and *Polygonum hydropiper* generally grow on the banks of lakes and by the side of drains, etc., while *Oldenlandia corymbosa* is a weed in paddy fields. *I. caruii* had high DM (17%) and poor extractability (7%). The N content and protein N extractability of *P. hydropiper* and *O. corymbosa* were large enough to make them hopeful plants for leaf protein extraction.

(C) *Plants giving a slimy pulp.* Juice could be extracted from *Corchorus olitorius, C. capsularis* (Jute), *Hibiscus sabdariffa* (Roselle) and *Ipomoea*

TABLE 3. Some species examined in West Bengal

| Latin name | Stage of growth | DM % on pulp | N % on DM of pulp | N % on DM of fibre | pH of extract | Total extractable N as % of total N in pulp | Extractable protein N as % of total N in pulp | % N on dry crude protein |
|---|---|---|---|---|---|---|---|---|
| *Eichhornia crassipes* | Flowering stage | 8·0 | 3·3 | 3·2 | — | 12 | 6 | 7·6 |
| *Pistia stratiotes* | Pre-flowering stage | 6·2 | 1·8 | 1·5 | 6·9 | 24 | 18 | 6·0 |
| *Ipomoea reptans* | Pre-flowering stage | 9·9 | 3·35 | 2·9 | 5·4 | 32 | 25 | 9·8 |
| *Polygonum hydropiper* | Pre-flowering stage | 12·8 | 3·2 | 2·6 | 5·4 | 33 | 23 | 8·0 |
| *Oldenlandia corymbosa* | Vegetative stage | 8·25 | 4·9 | 3·95 | 7·9 | 42 | 37 | 9·1 |
| *Hibiscus sabdariffa* | 53 days | 13·75 | 1·7 | 1·7 | 2·6 | 30 | 19 | 6·6 |
| *Ipomoea batatas* | Pre-flowering stage | 13·0 | 2·2 | 3·1 | 6·0 | 25 | 17 | 10·5 |
| *Pennisetum pedicellatum* | 36 days | 8·5 | 0·9 | 0·9 | 4·3 | 22 | 11 | 7·9 |
| *Pennisetum polystachyon* | Vegetative stage | 9·0 | 1·45 | 1·4 | 4·0 | 19 | 9 | 5·0 |
| *Solanum nigrum* | Flowering stage | 12·6 | 3·5 | 2·9 | — | 51 | 43 | 3·3 |
| *Cestrum diurnum* | Flowering stage | 16·8 | 4·3 | 3·7 | — | 41 | 32 | 9·2 |
| *Chrozophora plicata* | Seed stage | 17·4 | 2·8 | 2·35 | 7·9 | 38 | 24 | 13·6 |
| *Croton sparsiflorus* | Flowering stage | 18·3 | 2·95 | 2·5 | 5·7 | 40 | 32 | 8·3 |
| *Amaranthus gangeticus* | Flowering stage | 11·5 | 3·1 | 2·4 | 6·1 | 43 | 32 | 8·5 |
| *Datura metel* | Flowering stage | 11·5 | 4·25 | 3·95 | 5·0 | 26 | 12 | 7·2 |
| *Heliotropium indicum* | Flowering stage | 13·7 | 3·7 | 2·5 | 6·6 | 43 | 31 | 8·3 |
| *Alocasia indica* | Vegetative stage | 11·7 | 3·0 | 2·1 | 5·4 | 49 | 44 | 9·8 |
| *Mikania scandens* | Pre-flowering stage | 14·8 | 2·3 | 2·4 | — | 27 | 17 | 7·3 |
| *Helianthus annuus* | Post-flowering stage | 14·25 | 2·2 | 2·2 | 7·0 | 29 | 23 | 8·7 |
| *Pachyrhizus angulatus* | Pre-flowering stage | 12·85 | 3·3 | 2·5 | 6·0 | 40 | 25 | 5·2 |

*batatas,* in spite of the sticky texture of their pulps, but extraction from *Trillium* sp. was impossible. *Smilax* extracted well, but it is not common and the pulp contains a skin irritant. *C. olitorius, C. capsularis, H. sabdariffa* and *I. batatas* are by-products that are at present not used, or incompletely used. *H. sabdariffa* is unusually acid and would probably extract better if pulped with alkali.

(D) *Grasses.* A remarkable feature in the two grass species *Pennisetum polystachyon* and *Pennisetum pedicellatum* was that TCA soluble and TCA insoluble N, in the extract, were almost equal. Though the extraction of protein N was poor, they deserve further study because they re-grow well after cutting and *P. polystachyon* is perennial. Extraction may be bettered by use of alkali as both species gave extracts with low pH.

(E) *Wild plants. Solanum nigrum, Cestrum diurnum, Amaranthus gangeticus Chrozophora plicata, Croton sparsiflorus, Heliotropium indicum* and *Datura metel* extracted well; *Jatropha* sp. and *Lantana camara* did not. Some of these plants deserve further trial using samples grown with fertilizer.

(F) *Cultivated plants. Mikania scandens* is used as a cover crop. It grows luxuriantly, yielding approximately 10,000 kg/ha (fresh weight). *Alocasia indica* and *Pachyrhizus angulatus* leaves extract fairly well and are by-products. *Helianthus annus* grows luxuriantly and will be useful if protein extraction can be improved.

Thorough agronomic observations have been started on one cereal, bajra (*Pennisetum typhoideum*); one fodder grass, hybrid napier (*Pennisetum purpureum* × *P. typhoideum*); and three legume crops, dhancha (*Sesbania aculeata*), arhar (*Cajanus cajan*) and tetrakalai (*Phaseolus* sp.). Except bajra and arhar, the other three crops are not cultivated for human consumption. Hybrid napier is a good perennial forage grass, *S. aculeata* is cultivated as green manure and tetrakalai is a newly introduced fodder legume from Mexico. These are being grown on 4 m × 4 m plots from which 3 $m^2$ areas are being harvested after various periods of growth, and with varied seed rates and fertilizer.

The very large yields of leaf protein that are attainable was stressed in the descriptions of work at Aurangabad and in Ceylon and New Zealand. The work done at Calcutta and Mysore contrasted the yields from well fertilized and tended land with those from land getting little or no careful attention.

In Nigeria, peasant farming is still the rule. The farmer grows enough food for himself with some to spare for sale. When the land is exhausted he shifts to another area and allows the first one to revert to bush. By this shifting system of agriculture the farmer dispenses with fertilizer, though his yields are small by western standards.

Because of this factor, the first phase of work by Oke (unpublished) was concentrated on agronomic studies without the use of fertilizers. This will give an idea of the type of yield that would be expected locally.

Although all the results were obtained using the IBP pulper and press, work has started on the adaptation of the Posho Mill. The operation of this mill is similar to that of the IBP machine (Davys & Pirie, 1969). The mill is available in every village in Nigeria and is used mainly for grinding corn and beans. Since electricity has not reached all the villages, the mills are designed to run on diesel engines. In a comparison of the two machines using *Vernonia amygdalina,* the IBP pulper extracted 39% of the N and the Posho mill 35%.

The plants used are divided into three categories: cereals, legumes and green vegetables.

*Cereals.* The percentage N in the dry matter of maize increased from 3·0 to 3·5 in the first 5 weeks, and then started to decrease gradually to 2·5 in the eleventh week. The true protein followed the same pattern. The percentage dry matter increased progressively from 18·1 to 24·0 in 11 weeks. The amount of protein extracted ranged from 35 to 50%.

*Legumes.* Unlike cereals, in legumes the percentage N in the dry matter, and the protein N, increased up to 8 weeks before it started to decline. The total N increased from about 120 to 648 kg/ha. Over 70% of this was extractable during the first 8 weeks. Extractability decreased to about 20% after 11 weeks. About 80% of the total N extracted within the first 7 weeks was protein N and this decreased to about 30% after 11 weeks.

*Green vegetables.* As with legumes, the percentage N in the dry matter, and the protein N, increased for 7 weeks and then dropped more steeply than with legumes. The total N started to decrease after about 9 weeks. Up to 8 weeks about 90% of the N was extractable and 80–90% of this was true protein. After 11 weeks the true protein decreased to 40–50% of the N in the extract.

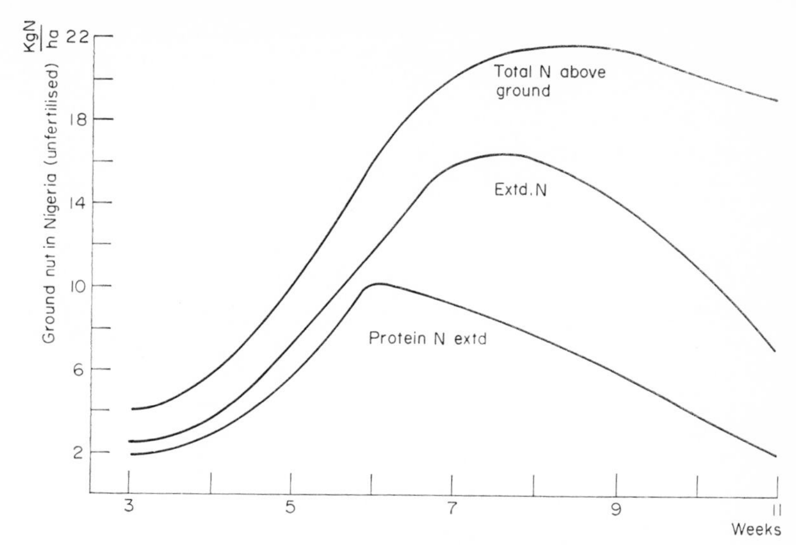


**Figure 3.2.** Ground nut in Nigeria (unfertilized).

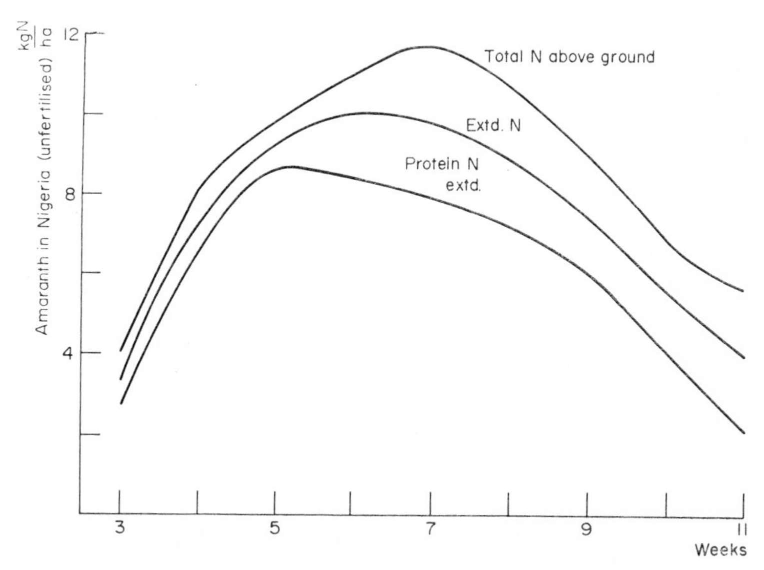


**Figure 3.3.** Amaranth in Nigeria (unfertilized).

The contrast between the behaviour of ground nut and amaranth is shown in Figs. 2 and 3. In spite of the diminution in the amount of N above ground after the 7th week, more protein can be prepared from amaranth than groundnut because protein extractability is not so strongly affected by age. This is a vivid demonstration of the point made by Arkcoll (p. 11) that careful agronomic work is needed to find the optimal time at which each species should be harvested.

Growth regulators, as their name implies, affect both the shape of a treated plant and also the amount of protein in it. An unsuccessful attempt was made (Byers & Jenkins, 1961) to increase, by treatment with gibberellic acid, the amount of protein in the regrowth from tares that had been harvested once. This experiment was done because it had been claimed that any increase in the first harvest, caused by gibberellic acid, was compensated for by a diminution in the regrowth. Since that time there have been several publications on increases in the percentage of protein N in the leaves of plants treated with the triazines, and the possibility has been mooted that this treatment could increase the yield of extractable protein. An increase in the concentration of leaf protein has been confirmed with *Phaseolus vulgaris, Pisum sativum* and *Zea mays* treated with 'simazine' and some other triazines. But there is no evidence that the amount of protein per plant is increased, rather than that the plants are stunted so that the same amount of protein appeared in less leaf, nor is there evidence that the extra protein (if any) is extractable. The effect of these and other growth regulators deserves fuller study.

Several of the species discussed by Joshi (p. 19), and by the authors quoted in this chapter, are at present classified as weeds. These plants are interesting because, although they do not produce a useful seed or tuber, or leaves that are edible in the untreated state, they could be useful if properly cultivated and if productive strains were selected. The erroneous suggestion sometimes appears, in popular accounts of work on leaf protein, that the untended growth on waste land should be used. That growth would be too sparse, illnourished and mixed to be useful. Water weeds are an exception to that generalization. Each area of water tends to be dominated by a single species and the water often contains nutrients such as NPK sufficient to encourage luxuriant growth.

Most of the papers published before 1967 on the use of water weeds are reprinted by Little (1968). Some other points on protein extraction and on the other uses to which water weeds could be put are made by Pirie (1970b).

Although many analyses have been published, they are often not accompanied by an adequate description of the physiological state of the material analysed. Table 4 gives some representative analyses taken from many sources; it shows the wide variation in N content that has been found for the same species and also shows that, when harvested at suitable times, some water weeds contain more N than conventional crop plants. This theme is dealt with in more detail in the next chapter.

TABLE 4. The nitrogen content of some water weeds

| Species | Popular name | Percentage of nitrogen in the dry matter |
|---|---|---|
| *Alternanthera philoxeroides* | Alligator weed | 1·3–3·5 |
| *Ceratophyllum demersum* | Coontail | 2·7–3·0 |
| *Eichhornia crassipes* | Water hyacinth | 1·3–3·7 |
| | leaves only | 5·0 |
| *Elodea canadensis* | Canadian pondweed | 2·2–6·3 |
| *Ipomoea reptans* | Water bind-weed | 4·6–5·8 |
| *Justicia americana* | Water willow | 1·6–3·8 |
| *Lemna minor* | Duck weed | 2·5–5·0 |
| *Myriophyllum spicatum* | Milfoil | 1·8–4·1 |
| *Phragmites communis* | Reed | 1·8 |
| *Pistia stratiotes* | Water lettuce | 1·7–3·9 |
| *Potamogeton* sp. | 'Pondweed' | 1·1–3·5 |
| *Salvinia auriculata* | Salvinia | 0·8–1·8 |
| *Typha* sp. | Reed-mace | 0·5–2·4 |

# 4

# Leaf Protein from Aquatic Plants*

C. E. Boyd

Noxious growths of aquatic plants often interfere with beneficial uses of freshwater lakes and streams in many regions (Holm *et al*, 1969). Millions of dollars are expended annually for aquatic weed control. Nutrient pollution is intensifying aquatic weed problems at an exponential rate by increasing the fertility of our waters. Control efforts are not keeping pace. Fortunately, there is a positive side to the aquatic plant problem. Polluted waters and even some uncontaminated bodies of water produce dense populations of plants of high nutritive value (Boyd, 1968, 1969a, b). Many aquatic plants should therefore be regarded as potential crop plants rather than biological nuisances. Food production by freshwater ecosystems would be greatly increased if leaf protein or even fodder could be made from water plants.

## Aquatic plant management

Aquatic plant problems generally result from one or more of four causes. (1) Plants often thrive in shallow, man-made impoundments and ditches where a large proportion of the system is suitable. Improper design may sometimes be blamed for infestations. However, regardless of water depth, any lake that is constructed in the tropics will likely become infested with floating aquatic plants. (2) Increased fertility (pollution) of aquatic habitats removes the nutrient limitations that were originally imposed upon the flora, thus resulting in more plant growth. (3) Introduction of species into areas where they did not originally occur often results in rampant growth, e.g. introduction of *Eichhornia crassipes* (water hyacinth) in the southern United States, Africa and India. (4) Some bodies of water that are unaltered by human activity contain dense stands of plants. During the natural process of

* Manuscript preparation was aided by Contract AT (38–1)–310 between the University of Georgia and the U.S. Atomic Energy Commission.

ageing (eutrophication) a lake that has been relatively free of vascular plants may suddenly develop weed problems.

Most aquatic weed abatement programs rely heavily on chemical control. Herbicidal applications are relatively effective in killing many species (Holm *et al*, 1969). This is only a temporary solution since plant growth usually appears again shortly after treatment and herbicides must be applied continually for satisfactory results. In addition, the long-term effects of herbicides may harm the aquatic environment (Hasler, 1969). Biological control may prove effective for some aquatic plants, but much additional research will be required to develop this technique. Inevitably, aquatic plant management procedures must be developed. Such techniques must be based upon sound ecological principles. We must recognize that vascular aquatic plants are a natural component of aquatic systems and cannot be eliminated from waters that are suitable for their growth. Elimination of a problem species from a body of water will lead to the development of a population of one or more other species (either vascular plants or algae) that is often as troublesome as the original infestation (Boyd, 1970a). Vegetation is likely to use existing space to a degree determined by environmental factors such as nutrients, light, bottom soil characteristics, flow, etc. We must learn to maintain species that are relatively innocuous to man's welfare. Locally troublesome growth can be controlled by chemical, biological, or mechanical means. Mechanical removal of vegetation is a particularly attractive method of control (Livermore & Wunderlich, 1969). This method does not appreciably alter the environment and plant-bound nutrients are removed with the vegetation. Plant harvest from effluent holding ponds is a potential method of stripping nutrients from effluents prior to release into the natural environment (Boyd, 1970a). Harvested plants could be used for leaf protein extraction or fodder. The worth of the plants as food would improve the economics of nutrient removal.

In tropical areas with food shortages, aquatic plants such as water hyacinth often represent a large food resource. The development of these weeds as crops for leaf protein production is logical and can be pursued independently of pollution abatement programs.

## Protein extractability and aquatic plant productivity

Aquatic and semiaquatic plants have a variety of forms; i.e. floating, submerged, floating leafed and emergent. The morphology and growth habits of a species influences its suitability for harvest and protein extraction. Differences in productivity and chemical composition of the different life forms are also pertinent.

For the ensuing discussion, standing crop data and the N content of plants will be reported on a dry weight basis. N will be used as an estimate of relative protein content although crude protein (N × 6·25) overestimates the true protein content of most aquatic plants by about 20% of the actual value (Boyd, 1970b). However, true protein does increase with increasing N content. The convention (N × 6·25) will be used to estimate leaf protein yields since the values agree closely with the true protein content calculated from the sum of the amino acids (Boyd, 1968).

*Eichhornia crassipes* and *Pistia stratiotes* (water lettuce) are the best known floating species. Westlake (1963) and Penfound (1956) and Penfound & Earle (1948) discussed water hyacinth productivity. Standing crops of 10,000 kg/ha are apparently common. However, the plants grow rapidly, and under a continuous cropping system yields in excess of 100,000 kg/ha appear attainable. *Azolla* and *Lemna* are widely distributed floating plants that produce small standing crops but grow rapidly and are suitable for continuous harvest. Floating plants should be relatively easy to harvest either mechanically or manually. Removal would not injure unharvested plants. Harvest and growth could be adjusted where continuous cropping is desirable.

Floating species have a high moisture content (92–97%) and contain 3–4% nitrogen (Boyd, 1968). Unfortunately, I have not been able to extract much protein from floating plants. About 15% of the total nitrogen in water hyacinth was recovered in leaf protein. Nevertheless, the extractability of protein from water hyacinth and other floating species should be thoroughly investigated since these species are abundant in most tropical countries.

Submerged species produce smaller standing crops than floating or emergent species; values greater than 5000 kg/ha are seldom encountered. The depth to which these plants grow depends on light penetration, but in most lakes submerged plants can be found to a depth of 2 m or more. Underwater cutting equipment required for mechanical harvesting is expensive and manual harvesting is rather difficult.

Submerged plants contain 90% or more moisture, and N levels range from 2 to 4% (Boyd, 1968). Less than 30% of the total N of several species was extractable as leaf protein. Poor protein recovery was due to large quantities of NPN in the extracts. Due to poor extractabilities, small standing crops, and difficulties in harvesting, the utilization of submerged species is questionable.

Floating leafed plants also produce fairly small standing crops (Boyd, 1968). Values for some common species are: *Nymphaea odorata* 1800 kg/ha, *Brasenia schreberi* 790 kg/ha and *Nelumbo lutea* 990 kg/ha. Floating leafed plants are typically found in water up to 2 m deep. Moisture and nitrogen contents are usually similar to those of submerged plants. Harvesting techniques used for submerged plants would be satisfactory for floating leafed plants.

Protein was readily extractable from *Nymphaea odorata* and *Brasenia schreberi,* but yields for *Nuphar advena* and *Nelumbo lutea* were small (Boyd, 1968). Protein yield per unit area would be low even for species that contain a high percentage of extractable N. Floating leafed plants hold little promise as a leaf protein crop.

Emergent species grow along the margins of lakes and streams to a depth of about 1·5 m. Above-water portions of emergent plants would be easy to harvest by hand. Mechanical harvesters to clip aerial shoots would not be difficult to construct. The moisture content of emergent plants ranges from 75 to 85%. N values vary greatly between species, but most species contain smaller percentages than plants with other growth habits. Morphological types found in emergent plants are similar to those represented by forage species, so protein extraction procedures for crop plants should be directly applicable.

*Justicia americana* and *Alternanthera philoxeroides* had peak total standing crops of 24,580 and 7980 kg/ha, respectively (Boyd, 1969b). Reed swamps often contain very productive emergent species. Standing crop values above 20,000 kg/ha were reported for *Typha latifolia, Arundo donax, Scirpus lacustris* and *Cyperus papyrus* (Boyd & Hess, 1970; McNaughton, 1966; Westlake, 1963). Standing crops of above-water portions will vary with the growth habits of the different species. Within a particular species, the proportion of aerial shoots will depend upon water depth.

Fairly good protein extractability (25–54% of the total N) was obtained for *Justicia americana, Alternanthera philoxeroides, Sagittaria latifolia* and *Orontium aquaticum* (Boyd, 1968). Other emergent species, *Typha latifolia, Scirpus* sp. and *Polygonum* sp. were unsuitable for protein extraction. Maximum protein yields for *J. americana, A. philoxeroides* and *S. latifolia* were 590, 478 and 362 kg/ha, respectively (Boyd, 1968). These values are as high as those reported for leaf protein yields of crop plants (Byers & Sturrock, 1965) and emphasize the value of aquatic plants as a crop. Swamps and shallow lakes in many regions often contain hundreds of hectares of emergent plants. Protein from these natural crops could supplement traditional protein supplies in many localities and alleviate nutritional inadequacies.

Of twenty-two species of aquatic plants (all life forms represented) that I have used in protein extraction trials, satisfactory results were obtained for six species. The ratio of satisfactory to unsatisfactory plants for protein extraction may seem to be smaller for aquatic than agronomic plants (Byers & Sturrock, 1965), but crop plants were chosen for digestibility and high nutritive content. Characteristics making a particular species suitable for forage also favour protein extractability. The percentage of native plants that make suitable cultivated forage crops is probably also very low. Only a few aquatic species have been screened for protein extractability and many species that will yield large quantities of protein undoubtedly remain to be found.

Species that cannot be used for leaf protein preparation can be used for fodder. The fibrous residue remaining after protein extraction can also be used as a low protein roughage. This residue accounts for roughly 50% of the original dry matter of aquatic plants.

## Resource development

In technologically advanced societies, utilization of aquatic plants for nutrient removal has great potential (Boyd, 1970a; Yount & Crossman, 1970). Food value of the plants is of secondary importance. Suggestions for pilot studies of nutrient removal and utilization of harvested plants are considered elsewhere (Boyd, 1970a). The food value of aquatic plants is of primary importance in tropical countries, but many tropical waters will benefit if nutrients are removed by plant harvest.

An approach possibly useful in developing aquatic plants into a crop for leaf protein extraction is presented in Table 1. Local aquatic species can be screened for protein extractability by small-scale procedures (Byers, 1961; Pirie, 1968b). Standing crop (see Westlake (1969) for techniques) and the available acreage of promising species should also be determined. Initial screening requires relatively little specialized equipment and is inexpensive.

Once potentially valuable species are selected, pilot studies to test large-scale extractability must be established (Davys & Pirie, 1963; Morrison & Pirie, 1961). The nutritive value of the leaf protein and its acceptance as a protein supplement in human diets must receive attention prior to continuous bulk production. Institutes that are working on leaf protein should include aquatic species in the program so as to prevent duplication of facilities. In addition, cost analyses must ascertain if it is economically feasible to produce aquatic plant leaf protein.

While bulk production is tested, field studies should be made of large-scale harvesting. Mechanical techniques may be adaptable in some regions, but, at least initially, manual harvesting will undoubtedly be the most effective. Timing of harvest to obtain maximum protein yield must be determined. Maximum leaf protein yield for *Justicia americana* was obtained 2·5 months before peak dry matter standing crop (Boyd, 1968).

Promising species are a crop; their destruction must therefore be avoided. Harvesting techniques which allow regrowth or methods of propagation will allow future utilization. Solutions to the problems of sustained harvest of a particular plant population will require ecological and physiological information on that species.

Exploitation of aquatic plants for leaf protein will present unique problems in each locality, depending upon species availability, harvest technique, environmental conditions and various undetermined factors. Therefore, development of this resource must be on a regional or local basis. In addition, it may often be more practical to test various species by trial and error rather than follow an elaborate experimental approach as outlined above.

TABLE 1. Development of a leaf protein resource from aquatic plants

| | |
|---|---|
| I. | Small-scale screening of local species |
| | a. Extractability of protein |
| | b. Yield of protein per unit area |
| | c. Local abundance of promising species |
| IIA. | Large-scale extraction trials |
| IIB. | Research on harvesting |
| | a. Harvesting equipment |
| | b. Timing of harvest for maximum protein yield |
| | c. Regrowth |
| | d. Propagation |
| | e. Ecological studies of promising species |
| III. | Nutritive value and acceptance of protein |
| IV. | Cost analysis |
| V. | Bulk production |

# Section II
# Extraction and Processing

# 5

# Equipment and Methods for Extracting and Separating Protein

N. W. PIRIE

## Equipment

Ideally, one machine would be used to pulp the crop and separate the protein-containing juice from the fibre. After testing many designs of screw-expeller and sugar cane rolls, we decided against them. The former were unsatisfactory because they consumed a great deal of power and, because of the continued rubbing in them, introduced an undesirable amount of finely divided fibre into the juice. Power consumption must be kept small, both for economy and to avoid over-heating the material in the machine. Unless a wasteful cooling system is used, it is clear from the 'mechanical equivalent of heat' that 50 HP for 1 ton (wet weight) of crop per hour is the limit in a cool country and a smaller value in a hot one. Several passes through rollers are needed to get satisfactory liberation of juice. This means, in effect, that several machines are being used even if they are all of the same type; the appearance of doing the job in one machine is therefore illusory. Davys & Pirie (1963) made an extractor that worked on 100–200 kg batches of crop; the process took 0·5–1 hour and, largely because of this prolonged exposure of juice to pulped fibre (Davys *et al,* 1969) the yields of extracted protein are only about half those got with more efficient equipment. Several of these 'village units' have been made, and they gave useful experimental quantities of protein, but we would not advise the manufacture of any more of that general design. If a small extractor is needed that is sufficiently slow-moving for it to be driven by an animal, its working volume should contain only 1 or 2 kg of crop. A probable design is a pair of perforated and ribbed coaxial cones with angles of about 30° and 20°. The inner (20°) one would be driven.

Having decided that pulping and pressing could probably not be managed on a large scale in one unit, we made a series of pulpers able to handle 1 ton of crop per hour. In the conventional fixed-hammer mill, the charge stays inside until it is sufficiently comminuted to pass through a grid. Pulp from a crop containing 80–93% water is an intractable mass that clogs all the grids tried. Clogging can be overcome either by drying the material so that it will blow through, or wetting it so that it will flow through. Drying is ruled out because it coagulates the protein on the fibre, and we think wetting inadvisable because of the great dilution of the resulting extract. The use of a conventional hammer mill on crops suspended in several times their weight of water has, under the misleading label 'impulse rendering', been advocated by Chayen *et al* (1961).

The principle underlying our pulpers has remained unaltered since 1950 but details improve; a phase in its evolution has been published (Davys & Pirie, 1960). The most recent is a cylinder 0·9 m long and 0·6 m in diameter within which a set of beaters, fixed to an axial shaft, rotates. For very tough crops, some prongs can be inserted through the casing into the spaces between the beaters so as to break the flow and increase the amount of effective work done. The crop is fed in tangentially at one end and the pulp discharges tangentially at the other. It comes out whether it has been properly disintegrated or not; it is therefore almost impossible to choke the pulper. Control over the amount of disintegration is given by using beaters differing slightly in form, by the use or otherwise of the prongs, and by varying the speed. After a few trials, pulping conditions can be found that are suitable for every crop from which protein could be extracted. The pulper runs at speeds from 800 to 1700 rev/min and with a 25 HP motor can handle 1–2 tons (wet weight) per hour when fed with succulent crops. Very fibrous crops go through more slowly. When the pulper cannot cope with more than 0·4 tons an hour, the crop is likely to be so mature that it will contain little extractable protein.

Recent improvements are mainly concerned with the arrangement and spacing of the beaters. As a result, an average of 1 ton of crop is now pulped in an hour for the expenditure of 10–40 kWh and 55–75% of the protein is extracted. We now realize that there is too much empty space in the middle of our pulpers and, when another is made, it will be smaller and faster. More attention than hitherto will be given to ensuring an even feed. Ultimately, it would be reasonable to expect the same efficiency of pulping for half the present power consumption.

Much of the protein in extracts from pulped leaves is in chloroplast fragments and other particles that are visible under the microscope. If pressure is applied to the pulp in such a manner that a thick compacted fibre layer is formed, these particles are filtered off. The basic principles of press design are therefore: the pressed layer should not be more than 6 mm thick, pressure should not be applied suddenly, and it should be maintained for several seconds so as to allow time for the juice to run away. In compensation for these restrictions, there is no need to apply pressure of more than 2 kg/cm$^2$. That is sufficient to press out 90% of the juice that is extractable at very much greater pressures.

Having discussed various ways for meeting these requirements (Pirie, 1959b), we made a satisfactory press (Davys & Pirie, 1965). This is an endless belt of woven nylon coated with PVC, 6 m in circumference and 0·3 m wide. It is tensioned and passes round a pulley (32 cm diameter) with a cylindrical face of perforated metal. The pulley, in a similar press made in New Zealand, is grooved rather than perforated. It is claimed grooves allow the extract to escape equally well and are easier to clean than perforations. The pulp flies out from the pulper on to the inner face of the belt and is carried by it round the pulley so that it is pressed between the belt and the perforated face. The belt is driven at 2–4 m/min by a 1 HP motor connected to the perforated pulley by a reduction gear and chain drive. The same motor drives an auger which removes the pressed fibre from the exposed surface of the perforated pulley.

The juice is pressed through the perforations into the pulley and then runs out over its edges into a tray. The belt is kept taut by a 0·76 m diameter pulley which is forced outwards by two pairs of springs, each spring exerting 1·0 ton. The size of the perforated pulley is dictated by the flexibility of the belt, on the one hand, and the need to maintain adequate pressure per unit area, on the other. The size of the large pulley is dictated by the operational convenience of having the end of the pulper encompassed by the belt-press so that the stream of pulp comes squarely on to it.

A small amount of the fibre is pressed through the perforations in the pulley and some is sometimes pressed out sideways. The juice is therefore strained through 0·15 mm aperture gauze. Commercially available strainers could be used for this process; we use a slowly rotating cylinder of gauze, 1 m in diameter and 18 cm along the face. The gauze is mounted between two annular plates (internal diameter 0·8 m) and its surface has a sinusoidal wave. This ensures that the fibre falls off it during back-washing and also that the juice does not lie only in the chamber made by the gauze and annular plates, but is carried some way up the ascending side rather than all collecting as a pool at the bottom.

Quantitative work on the extraction of protein from different crops, subjected to different forms of husbandry, is not possible with samples weighing less than 50 kg if these large machines are used. Detailed agronomic studies necessitate the use of many small plots and much of this has been done with domestic meat mincers followed by squeezing the pulp by hand in a cloth. Domestic meat mincers, both hand and power operated, are simple and widely distributed. On some crops they give repeatable results, but on others the scroll does not pull the charge smoothly into the barrel of the machine. The charge then has to be pushed in; this introduces a variable factor. The fibre in some types of leaf gets tangled round the cutter and stops the flow of charge completely. Stops for disentanglement introduce more variability. These faults are less obtrusive when a two- rather than a four-bladed cutter is used, and they can be minimized by giving the scroll and cutter independent drives and so breaking up the compacted mass that sometimes forms because of the constant relative positions of the cutter and scroll. Hand pressing is equally variable. These unavoidable elements of uncertainty in results with existing small-scale equipment convinced us that statements about the extractability of protein from crop plants, grown in different institutes, would not be comparable until the pulp was made and pressed in a manner that was less influenced by the skill of the operator.

The UK:IBP approved, in 1966, research on the standardization of methods for the agronomic study of leaf protein (*IBP News,* **20,** 1969) and arranged a grant to make the necessary equipment and train people in its use. At the expense of either the UK:IBP or local IBP committees, this equipment is now in use in India (two units), New Zealand, Nigeria and Sweden (two units). IBP units are also in use in Brazil, Ceylon, Eire and Pakistan but work with them does not figure in IBP programs.

There would have been no advantage in making a pulper able to give consistent results, comparable to those with the large machine, on less than 2 kg of leaf because it is necessary to have samples of that size if the character of a crop is to be fairly represented. The output of a grinding mill or pulper that is not empty at the end of a run may not have the same composition as the material retained. The error that this introduces into analytical work becomes smaller the larger the amount of material passed through the machine. At the end of a run the IBP pulper contains 200–300 g of material. If the first 500 g of pulp is discarded, the composition of the retained material is the same as that of the discharged material, 2 kg of crop is therefore sufficient—but it is better to use 3 kg. The pulp produced closely resembles that coming from the large pulper and the percentage of the protein that is liberated into the juice is similar (Davys & Pirie, 1969).

The IBP pulper is a stepped drum 44 cm long, 27 cm in diameter at the feed end and 32 cm at the discharge end; the difference in diameter ensures an air flow in the direction of movement of the pulp. A rotor inside the drum carries fifty-eight fixed beaters with a 2 mm clearance from the drum. In the laboratory, the rotor is driven by a 5 HP motor; the speed and direction of rotation are so arranged that the pulper can also be mounted on a 'Landrover' and driven from the power-take-off. It can therefore be used in the field. For safety the pulper is fed through a 5·5 cm tube. With the help of a plunger so shaped that it cannot reach the rotor, feeding at up to 1·5 kg/min is possible. When used by operators who can be trusted to work safely, a wider entry tube can be fitted, thus permitting faster feeding. The outer drum is easily taken off so that all parts of the machine in contact with the crop can be cleaned.

The pulper is supplied with several pulleys so that it can be run at different speeds. For most purposes, 3500 rev/min is satisfactory and to ensure comparability in results it would be well if this speed were adopted as standard. There is, however, no advantage in running at such a speed that all the fibres are chopped short; on soft material it will therefore be interesting to have some measurements at smaller speeds. If pulping is inadequate, undamaged fragments of leaf will stand out as pale patches in the dark pulp.

When protein is being extracted on a large scale the pulp is pressed between a belt and perforated roller without differential movement or rearrangement of the charge during pressing. The pressed fibre is about 5 mm thick and it is subjected to 1·5–2·0 kg/cm$^2$. Pulp from the IBP pulper is homogeneous; there is therefore no need to work with a large sample of it. But because conditions at the edge of the mass of material that is being pressed differ from those in the body of the cake, the larger the area of the cake the better. In the IBP press, 900 g of pulp is spread evenly on a cloth on a square grooved platen 23 cm each way, the cloth is folded so as to make the area of pulp 450 cm$^2$ and another grooved platen is placed on top. The assembly is mounted with the grooves vertical and subjected to a pressure of 1 ton applied by means of a bell-crank with 40 to 1 ratio. These conditions closely resemble those in the large press and the dimensions are chosen so that the unit can be carried and, like the IBP pulper, can be put in a 'Landrover' and used in the field. We have published detailed descriptions of these two IBP units (Davys & Pirie, 1969; Davys *et al*, 1969).

A crop that has been cut cleanly without bruising the leaves does not, so far as is at present known, deteriorate in a few hours in a cool climate. Even in a hot country a delay of half an hour between harvesting and pulping is probably not harmful. Obviously, many different plots should not be harvested at one time and then processed in sequence so that the last may have been deteriorating for several hours. Equally obviously, the condition of a crop early in the morning and late in the afternoon is not the same—especially in a hot dry climate. Judgement is therefore necessary in agronomic work and the plots should be replicated so that similar ones can be taken at different times of day and be pulped after different amounts of delay. After pulping, speed is essential because protein immediately begins to coagulate on to the fibre and to autolyse in the extract. A delay of 2 hr at 24° between pulping and pressing diminished the yield of protein from clover to half (Davys *et al,* 1969); Tracey (1948) and Singh (1962) got similar diminution with wheat. Pulp should therefore be pressed within minutes of being made. The continuous operation of the large-scale equipment automatically ensures a delay of only a few seconds between pulping and pressing. In large-scale work a second extract is usually made because, in cool weather, it may contain half as much protein as the first. If second extracts are being made, they should be made quickly. With slight modification, an IBP pulper can be used to make second extracts because the main work of leaf disintegration has already been done and 5 HP is sufficient to cope with re-wetted fibre coming from the large pulper-and-press system even when that is taking a lush crop at 1 ton/hr.

Because of autolysis, there should be no delay in quantitative work during the operations of measuring the volume of juice coming from the 900 g of pulp in the IBP press, mixing it, taking a sample and precipitating it with an equal volume of 10% TCA. Thereafter delay is immaterial. A crop, or potentially harvestable wild growth, can therefore be pulped, pressed and sampled with the IBP equipment at any site accessible to a 'Landrover' and the analyses can be completed at leisure in the laboratory.

The IBP pulper was designed as an agronomic tool and our aim was to make a machine as small as was reasonable considering the heterogeneity of a crop and the extent of fractionation that accompanies pulping. Nevertheless, the standard model takes 1 kg/min, with a larger entry-tube it takes 2 kg/min and, when fed mechanically at a uniform rate with chaff-cut leaf, it can take as much as 6 kg/min. The ability to handle 360 kg of crop per hour makes this pulper a useful small-scale production unit. But the IBP press is totally unsuited for production. It is designed to give uniform results and, in the time taken to arrange it for pressing one 900 g sample, several such samples could have been pressed, inconsistently, by hand. We have therefore made a small belt-press suited to the size and output of the IBP pulper. A description has not yet been published: it is simply a scaled-down version of the larger belt press.

## Methods of separation

Unless leaves are pulped to an extent greatly in excess of what is needed to liberate 70–80% of the extractable protein from them, the strained juice will not contain a harmful amount of fibre. It will, however, contain dust and other materials that may contaminate the leaf surface; it is therefore advisable to wash the crop before pulping it. We do this by submerging the crop in a tank in which a slow circulation of water is maintained so that the floating leaves are pressed gently against the bottom of the elevator leading to the pulper. This promotes even feeding. The water that clings to the leaf surfaces is beneficial because, although we do not think that the small extra yield of protein that is given when a large amount of water is added during pulping compensates for the difficulties introduced by dilution, a little extra water increases the efficiency of pulping and extraction. The optimum water content of the pulp seems to be in the range 90–93%. The juice from some leaf species, especially when harvested in sunny weather, contains starch. This, obviously, is a useful foodstuff, but it can be centrifuged off (along with some chloroplast material) if a final product with the greatest possible protein content is wanted. If it is not removed it will separate out along with the protein.

Leaf proteins coagulate easily. Ageing for 1 to 2 days or acidification can be used, but the coagulum is much easier to filter off if it is made by heating and heat diminishes microbial contamination and inactivates many enzymes. Furthermore, Subba Rau & Singh (1970) report slower growth in rats fed on acid-coagulated protein compared to those fed on protein coagulated by heat. To get a dense easily filtered curd it is essential to heat the juice quickly. On a small scale, this can be done by heating a stirred pan of water to 80°C and then running cold juice into it at such a rate that the temperature never falls below 75°C. The pan has an overflow for the curdled juice; the rate at which cold juice can be added depends on the means by which the pan is heated. There is risk of charring if cold juice is put into the pan initially and filtration will be difficult if any of the juice is underheated.

For continuous running, steam injection is preferable. Steam from a boiler (low pressure is adequate) is injected across the bottom of a U tube of 4–6 cm internal diameter pipe, 30–40 cm high. Cold juice is run into one of the open ends so that it flows round the U, in the same direction as the steam jet, and the curdled juice comes out of a side tube let into the other leg of the U about half way up. There is a dial thermometer in this leg and the same precautions have to be observed to keep underheated juice from coming out of the side tube. The rate at which cold juice can be run in depends on the amount of steam available. Ahmed & Singh (1969) suggest injecting the steam up the lower leg of a ⊣ and using it to lift the curdled juice into the filtration system. If the juice is alkaline, as it will be from some Cucurbitaceae, or if it has been made alkaline to promote protein extraction, it should be taken to about pH 5 before coagulation.

If heat coagulation is done quickly, and if no incompletely heated juice gets through, filtration is quick and easy. A standard filter press and cloths can be used or, on a smaller scale, long filter 'stockings' of calico which are pressed when they no longer drip freely. Whether made in a filter press or 'stocking' the final cake should be hard. The importance of this can be stated quantitatively. Completely pressed protein will contain only 60% water as measured by drying in an oven. That water is residual leaf sap and carries with it the soluble components of the leaf; the concentration of these will be about 5% but may be more. The 400 g of dry matter in a kg of protein, pressed thus, will therefore contain $600 \times 5/100 = 30$ g of non-protein solids, i.e. 7·5%. If, on the other hand, this press-cake contains 80% of water (it would feel fairly hard even with this amount of water in it), 200 g of dry matter would include $800 \times 5/100 = 40$ g of non-protein solids, i.e. 20%.

The protein from a few species (e.g. maize and pea haulm) has an attractive taste and can be used unwashed. Usually the cake is suspended smoothly in water and acidified. There are several reasons for adding acid. The second filtration is difficult with protein at pH 5–6 but easy at 3–4. A moist mass of protein at pH 4 has the keeping qualities of cheese or pickles; even when moulds grow on it, it is safe from food-poisoning organisms. Those alkaloids and other poisonous components that may be present in the crop, or in a contaminating weed, will be more completely removed by acid solutions. So far as is known, all the alkaloids would be removed in this way. It is traditional to use hydrochloric acid, sulphuric acid is as good so long as it is lead-free, the use of tamarind juice has been suggested. Acetic or formic acids might be used because of their bacteriostatic action. If importance is being attached to that quality, it would be better to add these acids at the end. Most of the acid runs away along with the cations in the wash-water and so is wasted.

There is little difference in the buffering power of protein from different species. The press cake is weighed roughly and the necessary amount of 2N acid is stirred in. As a rule, 200 ml will be needed for each kg of moist cake. The buffering power of the protein in this range is so great that there is considerable latitude in the amount of acid that can be added. The suspension is then filtered and pressed again. As a routine with all batches, a sample of the wash-water coming away towards the end of the draining period should be taken and its colour, flavour and dry-matter content noted. If it is strongly flavoured or contains more than 1 g dry-matter per litre, the washing should be repeated. Following the same argument as before: a cake containing only 60% of water, if this contains 1 g/l of solids, would be contaminated with only 0·15% of material that should have been removed, whereas a cake with 80% water will be contaminated with 0·4%.

The method of separation outlined is essentially that described by Morrison & Pirie (1961). At that time we did not realize that there was any advantage in separating the curd quickly from the uncoagulable components of the juice. We suggested (Pirie, 1968a) that precipitation by phenolic compounds could explain the poor extraction of protein from some leaf species, and we know that unsaturated fats could make the extracted protein less digestible by enzymes (Pirie, 1966a; Buchanan, 1969a, b). The work of Pierpoint (1969a, b) and others made us wonder whether tannins, quinones and other such substances might form soluble complexes with leaf protein as well as coagulating it on to the fibre (Pirie, 1969a, b; Byers, 1971) and that some of the differences in nutritive value between different protein preparations arise from differences in the extent to which such complexes have been allowed to form. Protein that is processed quickly is more readily digested by enzymes, contains more available lysine, and less methionine sulphoxide than protein processed more slowly. More evidence on this point is given in the accompanying paper by Allison (p. 78) and by Tannenbaum *et al.* (1969).

Because of the probability that complexes will be formed, and because the soluble carbohydrates and other uncoagulable components of leaf extracts are either harmful or of little nutritive value for single-stomached animals, we disagree strongly with the suggestion (Hartman *et al*, 1967) that the unfractionated leaf extract should be dried, and with the suggestion by Hollo & Koch (p. 65) that it should be concentrated and mixed with the coagulated protein. Subba Rau *et al* (1969) have demonstrated, by feeding experiments with rats, the damage done by evaporating the whole extract.

Before leaf protein can gain acceptance as a human food, standards of quality will have to be drawn up. Meticulous care will be needed to meet these standards. Milk production offers a good analogy. The milking shed is essentially a dirty place, but the dairy is hygienic. Similarly, the crop, as harvested, is dirty. It needs careful rinsing to remove dust and other contaminants from its surface. Otherwise there will be an unacceptable amount of acid-insoluble ash in the final product. Conditions around the pulper and press will inevitably resemble those of the milking shed, the extract, however, should be treated with the care and cleanliness that characterizes a dairy.

The reasons given for extracting the protein with dilute acid outweigh the disadvantage that this extraction also removes much of the calcium, magnesium and potassium. These are useful components of a diet, but they are easily supplied in other ways. Coagulation at 80° kills most of the bacteria and other possibly harmful organisms on leaf surfaces. If this treatment should not, in some circumstances, be thought adequate, the final product may be given a more intense heat treatment, or hypochlorite or some such agent could be added to the water in the original rinsing tank or to the water used to wash the protein. These are matters for further experiment in the conditions prevailing at each place where leaf protein is being made. The acceptability of crops that have been treated with insecticides or herbicides is also a matter for experiment. Many are harmless and others are probably removed from the protein by washing. But until these points have been definitely established, protein from treated crops should not be used as human food.

# 6

# Commercial Production in Hungary

J. Hollo and L. Koch

We consider a basic change in human eating habits unlikely. With this in mind, crops harvested at a time when they are at the peak of their vegetative growth should be suitably processed and fed to animals. Such a scheme would avoid the inevitable losses when protein is translocated from leaves to seeds, in soyabeans, etc., which are, at present, used partly for feeding animals, but partly also for direct human consumption. The role of manufacturing industry would be to produce raw materials and additives which would give optimum yields. Industrial products gained through synthesis or fermentation processes, e.g. yeast grown on oil, seaweed, and synthetic amino acids, would be used indirectly in fodder. The raw materials from agricultural production would thus undergo reprocessing in future. We use not only of the complete protein fraction of the raw material, but also the soluble components of the plant. The high-fibre by-product could also be used with, for example, urea, in feeding polygastric animals.

## Technological problems

To achieve the above goals certain technical problems needed to be overcome. Of prime importance is that the product should be of even composition especially in regard to the ratio of calories to protein or amino acids, although the composition of raw materials to be processed will in many cases vary. A change in the composition pattern has to be allowed for, even if the species to be processed remains the same. Especially in the temperate zone, several species of plant are used instead of a single one. This is an important consideration, as the time factor of a given production process can become crucial in the redemption of investments.

The procedure used begins towards the middle of April, with processing rape either alone or in mixtures. The procedure ends in December or January with field kale. Thus an estimated processing time would be about 250 days even in the temperate zone. Such a procedure requires an adequate organization of production.

When isolation is not confined to the protein fraction alone, problems arise from the enrichment of accompanying substances—saponins and mustard oil, etc.—in the end product. When it is possible to use raw material of one or only a few plant cultures, e.g. in tropical regions, the technological problems are simpler. However, in the temperate zone, the raw materials involved extend over a very large number of plant species and the technical problems are increased.

### Process stages

1. Firstly the raw material is subjected to mechanical disintegration. In contrast to the methods aimed at the production of protein for human consumption, in this method no additional liquid (i.e. water) is used.

After disintegration, the plant pulp undergoes multi-stage pressing. In the course of this operation, the press cake is re-extracted with the supernatant made when the first extract is heated. A second disintegration is made only if a yield higher than 50% of extractable matter is required. In the final pressing stage, the residue from the low pressure process is treated in a high pressure unit—a screw-press best serves this purpose—in which the water content is reduced to 45–50%. The end-product of the last pressing stage is a highly fibrous residue. This substance itself or with urea is passed, pneumatically or by means of a conveyor belt, through a drying tunnel. The final product emerging is a coarse-grained mealy material, which is then packed directly, or after briquetting. The highly fibrous by-product has, without any artificial additives, 30% of the overall nutritive matter content of the raw material and is therefore suitable fodder for ruminants.

2. All the extracts are collected and after heating to 75–85° the suspension, together with the chloroplasts and protein fractions which have settled out, are separated in a continuous process. The separated sludge is green, and its solid matter content totals 25–30%.

3. The clear liquor yielded by separation passes to a buffer reservoir from which it is fed continuously into a vacuum evaporator. When entering the evaporator, the liquor has a solid content of 5–12%, depending upon the type of primary material processed; the concentrate contains 55–70% solid matter.

4. The concentrate is mixed with the separator sludge to give a pulp containing 40–50% DM. This is then fed into a spray-dryer. It is dried by flue gas at an inlet temperature of 250–300°. An Anhydro spray-dryer with direct oil firing is used. Once cooled to 30°, the powder thus obtained is loaded into sacks, or is pelleted without prior cooling and then cooled and packed.

Economy is dependent on two factors: (i) a sufficient yield of the main product; and (ii) the thermal efficiency of vacuum evaporators.

TABLE 1. The nutrient values of different crops harvested at their vegetative phase

| Plant | Crops in 100 kg/ha | | |
|---|---|---|---|
| | Dry matter | Raw protein | Essential amino acids |
| Soyabean seed | 14·4 | 5·1 | 2·3 |
| Fodder cereals | 35·0 | 4·2 | 0·8 |
| Lucerne | 45·0 | 9·8 | 3·8 |
| Various plant series harvested in vegetative stage | 84–105 | 31–40 | 12–16 |

## Protein quality

The main product contains most of the true protein of the plant. From 1 ton of raw green matter with a 20–22% dry matter content, the yield is 90–100 kg of the concentrate with a low fibre content, and 100–110 kg of by-products. Depending on the quality of the latter, it is possible to keep the specifications of the main product within certain limits. These are shown in Table 2.

TABLE 2. Specifications of main product from green matter

| | % |
|---|---|
| Raw fibre content | 1– 5 |
| Raw protein content | 40–44 |
| Crude fat content | 2– 4 |
| Ash | 12–20 |
| N-free extract | 40–45 |

The fact that the fibre content can be kept within limits is important. The end-product if it is to be of use as a substitute, in part, for powdered milk in rearing young domestic animals, in addition to having a low fibre content, must have a protein level of around 40–44%, irrespective of the primary material processed. From experiments with animals it has been found that the biological value, including amino acid content, of this food is as high as that of soyabean protein. The protein and the amino acid content is shown in Table 3.

TABLE 3. Amino acid content of the main product from green matter

| Amino acid | g/16 g N |
|---|---|
| Histidine | 2·4 – 2·6 |
| Lysine | 6·5 – 7·5 |
| Arginine | 6·0 – 7·6 |
| Cystine | 1·0 – 1·5 |
| Methionine | 1·7 – 2·2 |
| Phenylalanine | 4·8 – 5·6 |
| Tryptophan | 1·3 – 1·5 |
| Valine | 5·5 – 7·0 |
| Threonine | 5·3 – 5·8 |
| Glycine | 4·6 – 5·8 |
| Leucine-isoleucine | 11·4 – 13·4 |
| Tyrosine | 3·5 – 4·5 |
| Glutamic acid | 7·0 – 9·0 |
| Aspartic acid | 6·0 – 8·0 |
| Serine | 5·8 – 6·4 |
| Proline | 3·5 – 4·8 |
| Alanine | 4·2 – 5·8 |

The correct choice of the raw material should not be overlooked from the point of view of economy and potential end-product. Research is being carried out with plant species that were hitherto not utilized in agriculture. Species such as *Amaranthus, Atriplex* and *Chenopodium* show favourable qualities, e.g. *A. hybridus* has a very high lysine and methionine content.

In addition to the macro-components, the remainder of the constituents of importance in animal rearing and feeding should also be considered in evaluating the main product. These have primarily included vitamins, plant pigments and other factors that further growth. As the main product also contains the water soluble ingredients of the plant, it might be claimed to be complete in respect to the nutrient components of a plant.

The carotene and xanthophyll contents of the main product are high, around 300 mg/kg of β-carotene. It is also of interest from the point of view of biological value that, unlike other common green plant processing methods—where drying is effected in a hot air atmosphere using flue gas heating at a very high intake temperature level, this process requires a maximum temperature of only 105°. The biologically important constituents of the main product are shown in Table 4.

TABLE 4. Biologically important constituents of main product from green matter

| | mg/kg |
|---|---|
| Vitamin $B_1$ | 10– 17 |
| Vitamin $B_2$ | 18– 21 |
| Vitamin $B_6$ | 15– 18 |
| Niacine | 85–115 |
| Pantothenic acid | 35– 45 |
| Folic acid | 1·8–2·5 |
| Choline | 2200–2600 |
| β-Carotene | >300 |
| Xanthophyll | >600 |

## Energy requirements

An important factor that determines process economy is the cost of the energy source. A considerable amount of electrical energy is required in the disintegration procedures and in the separation of the raw material, in order to obtain two phases. This requirement for electricity is about twice that normally consumed in common drying procedures; however, because 70–80% of the water is evaporated in a vacuum evaporator it is possible to economize considerably in the heat consumption. In hot air lucerne drying equipment, the specific calorie consumption is as high as 900–1000 kcal to vaporize 1 kg of water, the same value for a multi-stage condenser, used in the author's process, is only 300–350 kcal.

The pilot plant with a daily production of 40–60 tons of green matter, furnishes ample evidence of this. The production figures show that 45–55 kWh of electricity and about 40 kg of fuel oil are required to process 1 ton green matter.

## Feeding trials

The end-products thus yielded have already been applied in *in vivo* experiments. The primary aim of the trials has been to reduce the soyabean and peanut content of mixed fodders. The production of the latter, in Hungary, has reached a considerable level, but all the necessary protein carriers, including soyabean, peanut meal and fish meal, are imported. The bulk of these mixed fodders, and therefore the major part of the imported protein carriers, are used in piggeries and poultry farms. Experiments have therefore aimed at replacing the soyabean and peanut meal extracts in the fodder of these domestic animals.

So far, the green plant concentrates, added to the fodders, have been used at up to 7·9%, without any regression in the specific transformation ratio of fodders into meat. In this way, a 25–30% reduction in the soyabean and peanut content of the experimental fodder mix has been achieved. Also an increase in the rate of live-weight gain of the experimental animal group was achieved, whereas the specific per-weight-unit fodder consumption is the same as that of the control group of animals. The goal is to bring the percentage of green matter concentrates in the fodder mix up to 15–20%, thereby replacing the protein carriers of vegetable origin by 50–60% and the fish meal quota by 20–25%.

When a green plant concentrate is used at a very high level, certain accompanying substances with a possibly unfavourable effect may influence the efficiency of the feed. It is possible by suitable choice of raw material and processing techniques to prevent the unfavourable substances from reaching undesirable concentrations in the main product.

In addition to these goals, experiments are also included on the replacement by green plant concentrates of powdered milk, up to a certain percentage, in the fodder of young animals. A further requirement is to produce fodders with high β-carotene and xanthophyll content for feeding laying hens.

The present unit takes 5 tons of crop per hour and a larger unit will be operating in 1972.

7

# Commercial Production from Alfalfa in USA

G. O. Kohler and E. M. Bickoff

An economically sound process for large-scale production of leaf protein must also yield high value feed products. Some research has been directed toward ensiling or making hay from the fibrous residue after extraction. Our approach is divided into two phases: Phase I—to develop a basic wet separation process to produce: (a) standard grade dehydrated alfalfa meal; (b) 50% protein, high-xanthophyll concentrate for poultry; and (c) forage solubles concentrate for use as an unidentified growth factor supplement for livestock (a full-scale plant to process 50–70 tons of alfalfa per hour is completed, based on our Phase I pilot plant); Phase II—a research program to develop further economically sound steps to yield pigment-free, palatable LPC plus xanthophyll concentrate from the intermediate or end products.

Grasses, alfalfa, and other forage plants constitute the world's largest usable supply of protein and provide the basis for animal agriculture. According to the FAO Yearbook (1968) there are over 2600 million hectares of meadows and permanent pastures in the world. In the USA alone, over 130 million tons of grass and legume hays are produced annually, of which 74 million tons is alfalfa hay (U.S. Stat. Rept. Ser., Crop Prod., 1969).

Production and utilization of forages have reached a high level of efficiency as the result of research by government and commercial organizations. A relatively sophisticated dehydration industry, from a technological standpoint, has evolved which has been growing at a steady rate on a world-wide basis. This forage dehydration industry sells its products, largely 17–20% protein alfalfa, at a retail price of 40–60 dollars/ton, F.O.B. plant or 24–36 cents/kg of protein. These prices include profits to the farmers and processors. The raw material, fresh forage in the field, sells for 10–20 dollars/dry ton or 6–12 cents/kg of protein. With soybean meal at 70 dollars/ton the soy protein costs about 24 cents/kg. Thus, forages provide a very low cost raw material for protein recovery.

From a nutritional standpoint it has been shown that the amino acids of alfalfa are well balanced and at least equivalent to those of soy protein.

It is little wonder then that scientists and inventors have been trying for more than 50 years to convert leaf products directly into human food. The key to such conversion lies in separating the protein from indigestible plant components such as cellulose, pentosans, etc.

Osborne & Wakeman (1920) stated in their classical paper on spinach leaf protein, 'If we can learn to separate the contents of the cell from these (e.g. cell walls and water) we shall obtain a food product of very great value'.

Considering the tremendous supply, the low raw material cost, and the amount of research done, the logical question is, 'Why hasn't a leaf protein concentrate been developed on a commercial scale?' We shall attempt to answer this question, considering first the technological problems, second the quality problems, and thirdly economic factors. We shall then describe the approach which we are pursuing to overcome the problems which, up until now, have provided an effective block to commercialization.

The basic steps for preparation of LPC have changed little during the past 30 years. As shown in Fig. 1, they consist of grinding the plant material to rupture the cells and pressing to separate the juice from the fibre. The juice, which contains soluble protein and suspended chloroplasts (Kohler & Graham, 1951), may be concentrated and/or dried without further separation (Hartman *et al,* 1967). Alternatively, it may be separated into a green protein coagulate (LPC) plus a clear brown coloured solution (alfalfa solubles) after treatment with heat, acid, solvents, etc. The coagulate LPC may be stabilized by adding salt, drying or other means and the soluble fraction may be stabilized by concentrating to 50% solids or higher (Bickoff *et al,* 1947; Chayen *et al,* 1961; Crook, 1946; Duckworth & Woodham, 1961; Henry & Ford, 1965; Lugg, 1939; Kohler & Graham, 1951; Morrison & Pirie, 1961).

While these steps are relatively simple in theory, their translation into a large-scale process presents a number of difficulties. In most cases percentage recoveries of the crude protein of the raw material have been low. In the first place only 70–75% of the N in the leaf is in the form of true protein. The remainder is non-protein nitrogen. While about 60% of the NPN is made up of free amino acids, they are not recoverable by the coagulation step. They would be included in a whole juice LPC, but, as has been pointed out by Pirie (1969b; 1970a), the soluble fraction of many leaves contains strongly flavoured substances and, in some cases, alkaloids or glycosides which may be nutritionally undesirable. For these reasons, we shall limit further discussion in

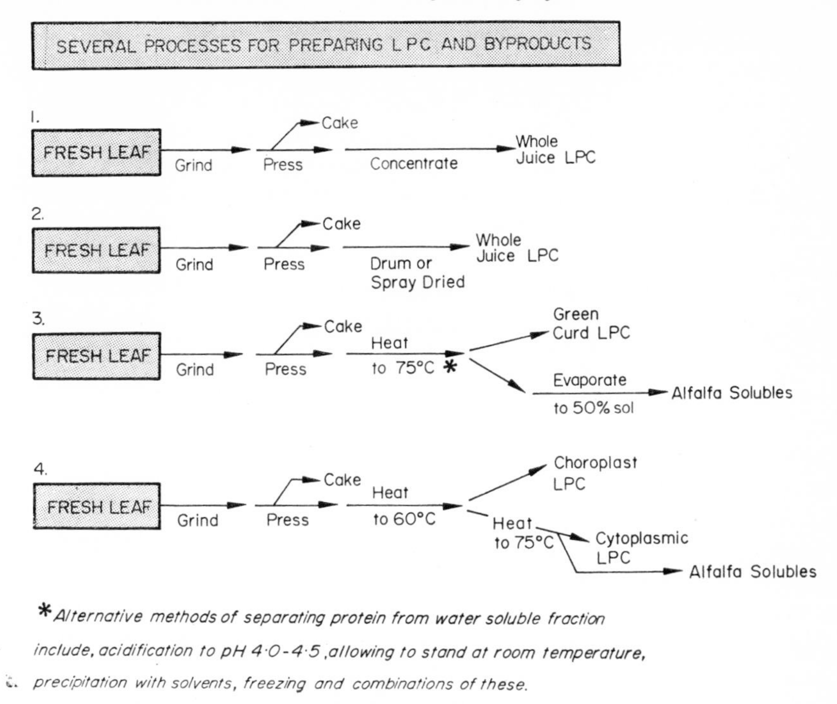


Figure 7.1. Several processes for preparing LPC and byproducts.

this presentation to LPC which has been coagulated and separated from at least the bulk of the leaf solubles fraction. Yields of this type of LPC must thus be calculated based on the coagulable protein rather than crude protein in the raw material.

A second factor which makes high yields difficult to attain is the fact that the maximum yield of protein attainable is limited by the number of cells torn open in the grinding process.

As Osborne & Wakeman (1920) pointed out, it is virtually impossible to rupture 100% of the cells on a large scale. From the maximum yields obtainable by various workers it seems likely that a figure of about 75% breakage of forage plant cells is close to the maximum achievable under practical conditions. Figures as high as 90–95% have been reported in laboratory studies on such low fibre plant leaves as tobacco (Crook, 1946) and spinach (Lugg, 1939).

A third factor which limits yields during pressing is the fact that the fibrous mat tends to act as a filter which holds back chloroplasts (Davys & Pirie, 1965). Indeed grinding chloroplasts with fibre actually causes adherence to the fibre (Crook, 1946). Since close to half of the true protein of the leaves is in the chloroplasts, filtration losses can be serious.

The design of the press should therefore minimize the thickness of the layer of material being pressed (Davys & Pirie, 1965; Morrison & Pirie, 1961). Pulping with an excess of water present (Chayen *et al*, 1961; Morrison & Pirie, 1961) and rewashing the press cake (Bickoff *et al*, 1947; Morrison & Pirie, 1961) are means of minimizing or overcoming the filtration effect.

So much for the technological problems involved in obtaining high yields of LPC. Let us consider next the properties of the products themselves. From an analytical standpoint, LPC contains a very desirable array of essential amino acids (Chibnall *et al*, 1963; Gerloff *et al*, 1965). Regardless of the leaf source the amino acid composition of the LPC is remarkably constant. The first limiting amino acid in all cases is methionine. LPC is highly susceptible to heat damage so that processing must be carefully controlled to avoid losses in biological value and in protein digestibility as determined by both *in vivo* and *in vitro* methods (Davies *et al*, 1952; Duckworth & Woodham, 1961; Henry & Ford, 1965). There is evidence that the lipid oxidation which occurs during drying at elevated temperatures is involved in the reduction in digestibility (Duckworth & Woodham, 1961). The Maillard reaction (e.g. interaction between reducing sugar and free amino groups) has also been implicated. The biological values (BV) of freeze-dried LPC preparations fall in the range of 70–80% while true digestibilities (TD) run about 75–85%. Oven drying at 100° dropped BV and TD values by 5–6% and 10–20%, respectively.

The heat coagulable protein from leaf juice can be separated into two fractions by differential coagulation (60° followed by coagulation at 80°). The 60° coagulum is largely 'chloroplastic' protein while the 80° coagulum is largely 'cytoplasmic' protein. The chloroplastic protein has a somewhat lower biological value and true digestibility than whole unseparated LPC. The cytoplasmic protein has proportionately higher values (Davies *et al*, 1952; Henry & Ford, 1965).

The overall assessment of nutritional value of carefully dried LPC is that the BV of its protein is superior to that of soybean and other seed proteins and approaches that of milk protein.

About 3 years ago we re-entered this field of research in collaboration with a company which produces dehydrated alfalfa in bulk. In contrast to the

PLATE 1

The IBP pulper open for cleaning and mounted for use in the laboratory.

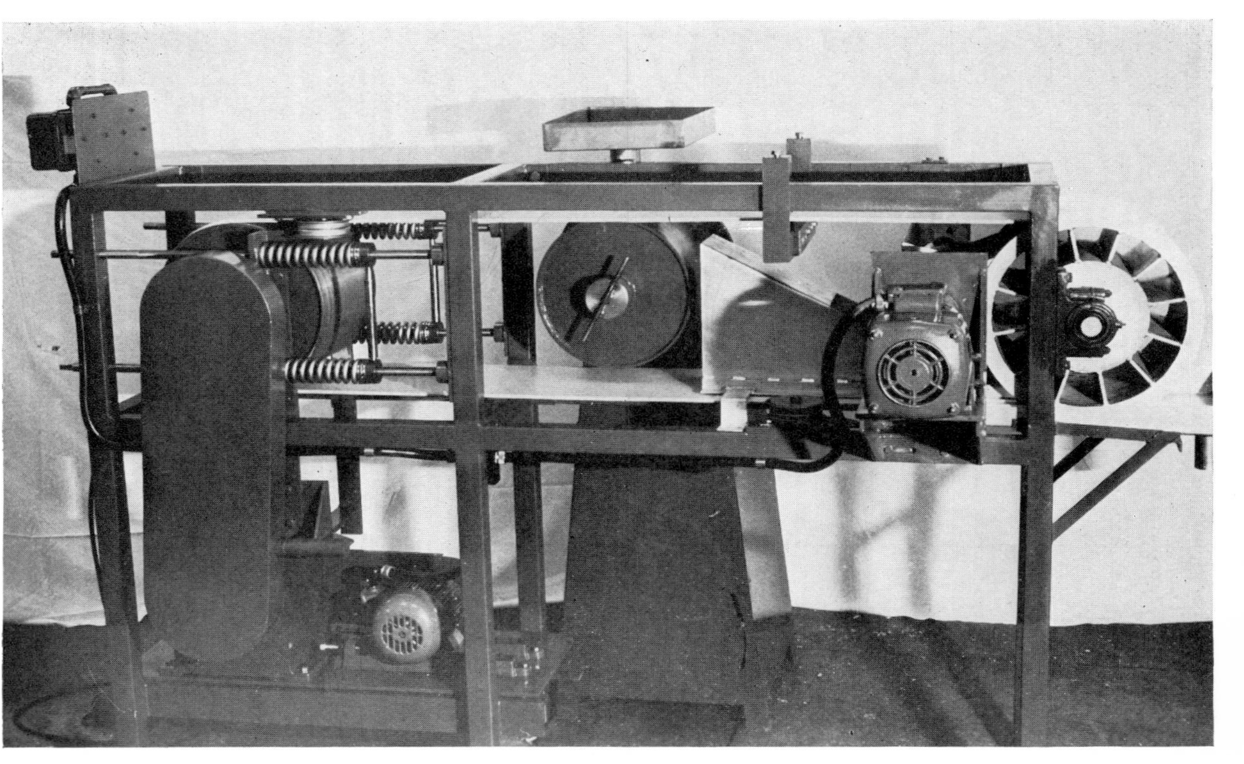

PLATE 2

The IBP pulper used in conjunction with a small belt-press for making protein from about 150 kg of crop per hour.

Plate 3

The IBP press designed for use in agronomic experiments. It is shown here with 10 kg applying pressure, to the platen at the right, and 15 kg ready to put on so as to reach the full 1 ton on the platen.

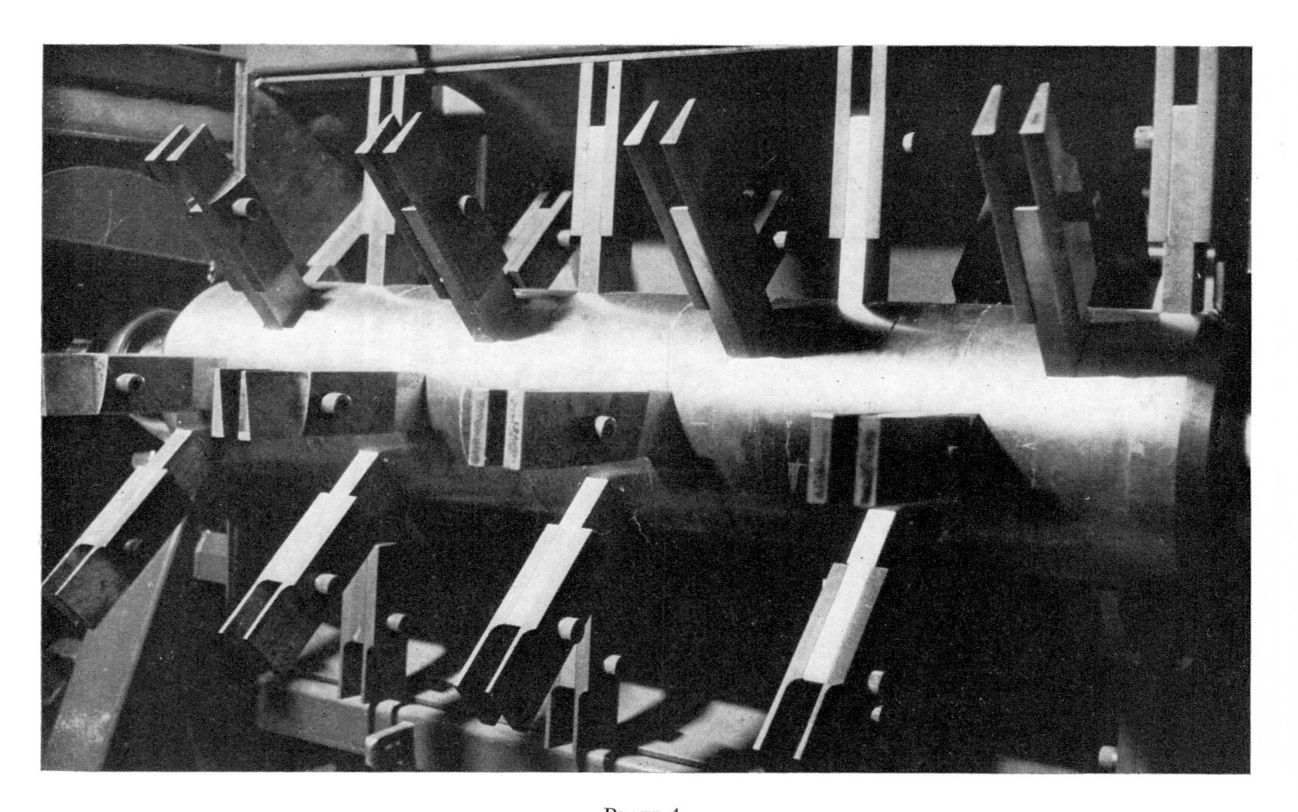

PLATE 4

The arrangement of beaters in the most recent pulper used at Rothamsted to take more than 1 ton of crop per hour.

PLATE 5

The 1-ton-an-hour arrangement at Rothamsted. Pulp flies out from the pulper in the middle on to the belt which carries it to the right and presses out the juice. This is strained, to remove traces of fibre, at the extreme right.

PLATE 6

The largest belt-press made so far. It is expected to take 3 tons per hour.

PLATE 7

The alfalfa processing unit in California. The conveyor in the foreground takes the crop to a roll-press and the conveyor in the middle is taking the fibre to a drier.

earlier work which focused on leaf protein for human food, we evolved a program, the immediate (Phase I) objectives of which were to produce superior animal feeds. We felt the success of this Phase I effort would serve as an economically sound base for a Phase II development of low-cost LPC.

The Phase I research objectives were: (a) reduction of costs of dehydrated alfalfa by mechanical dewatering; (b) recovery of a high-xanthophyll–high-protein product (PRO-XAN) designed primarily as a pigmentation supplement for poultry, and (c) recovery of the solubles fraction from the coagulation step as a molasses or unidentified growth factor concentrate. Phase II of the program would then be conversion of the protein-xanthophyll concentrate into a bland edible protein product and a stabilized highly concentrated xanthophyll product for poultry rations.

Preliminary reports on the Phase I process (PRO-XAN process) have appeared (Kohler *et al,* 1968; Spencer *et al,* 1970) and research on the unit processes are now being written up for publication. The process basically follows the LPC process mentioned above, added to the conventional dehydration process. However, the objectives of Phase I required that certain process changes be made.

Since we were not interested in maximum yields of protein but rather in large throughput and maintaining quality of the pressed cake to produce a high-grade alfalfa meal, we chose a sugar cane roll press to replace both grinding and pressing operations of the commonly used processes (Fig. 2). An important consideration was that sugar processing equipment is available for very large scale operations and would require little or no machine research and development. Roll presses had been tried by others and judged to be relatively inefficient (Morrison & Pirie, 1961). In our pilot plant rolls we can obtain 35–50% of the weight of the fresh young alfalfa as press juice at a feed rate of about 400 kg/hour. Fig. 3 shows a balance sheet on solids, protein and water in a 35% juice removal operation. Alfalfa which normally would give a 20% protein dehydrated product yielded a press residue of 19·6% protein. Table 1 shows that dehydrator throughput was increased by about 40% in a typical experiment, thus greatly reducing the cost of dehydrated alfalfa. The pressed dehydrated alfalfa could still be air-classified to produce a 25% protein meal if desired (Chrisman & Kohler, 1968) for use in rations of monogastric animals.

Initially we were dismayed to find that much xanthophyll was lost during and immediately following rolling. We found that by adjusting the pH of the alfalfa to 8 or over the losses of xanthophyll were reduced, presumably due

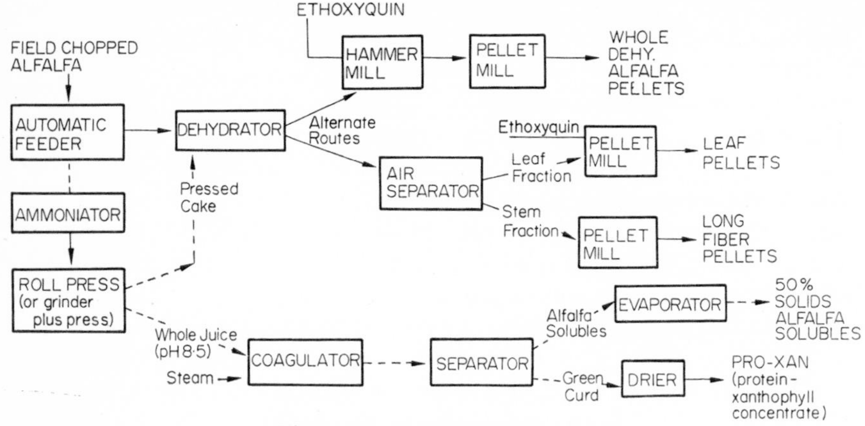


Figure 7.2. The Pro-xan process.

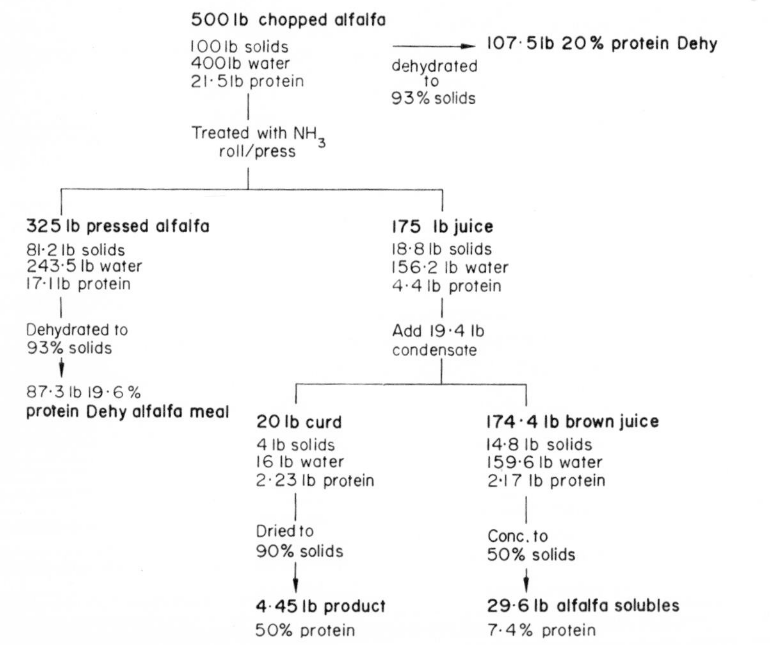


Figure 7.3. Material balance through pro-xan process at 35% juice removal level.

TABLE 1. Comparison of dehydration rates and separability of whole alfalfa and dewatered alfalfa

| | Dehydration rate (kg/hr) | % Leaf separated | Protein % on DM | Xanthophyll (mg/kg) |
|---|---|---|---|---|
| Whole alfalfa | 152 | — | 22·8 | 450 |
| Leaf fraction | — | 42·3 | 25·1 | 474 |
| Dewatered (dehy) alfalfa | 202 | — | 21·3 | 405 |
| Leaf fraction | — | 42·0 | 24·5 | 452 |

to the low activity of lipoxidase at more alkaline pHs. This brings us to the second change which has to be made. Going back to Fig. 2 we developed conditions to avoid xanthophyll losses in the PRO-XAN process. The importance of these changes is illustrated by the fact that computer evaluations of projected products show that the xanthophyll in a low fibre product is worth 10–20 cents/g depending on types of poultry rations being considered and prices of competitive xanthophyll containing feeds. Thus, a PRO-XAN with 40% protein and 1·7 g/kg xanthophyll would have been worth over 300 dollars per ton in a layer ration based on the Kansas City, December 1966 market (Taylor *et al,* 1968).

The addition of ammonia was found to have many beneficial effects (Table 2). It causes the curd formed on steam injection to be harder and more readily handled. It keeps the product green by preventing conversion of chlorophyll to pheophytin. This is desirable in a feed product. Although it

TABLE 2. Benefits of addition of ammonia at the grinding pressing step of PRO-XAN process

1. Increased recoveries of xanthophyll due to reduced lipoxidase activity
2. Maintenance of green colour (e.g. chlorophyll)
3. Improved curd hardness and handling properties
4. Increased yield of protein
5. Reduced autolyses of protein due to reduced protease activity
6. Ammonium salts in the alfalfa solubles byproduct are utilized as NPN for ruminants

has not been demonstrated we feel that the yield of protein should be increased since other bases have been shown to have such an effect (Crook, 1946; Datta *et al,* 1966; Nazir & Shah, 1966; Singh, 1964). And finally, the addition of ammonia would be expected to reduce protein losses through autolysis during handling and extraction. This is suggested by the work of

Singh (1962) who found that an enzyme system of wheat leaves shows maximal activity at about pH 5·5 and is less active at pH 8 and over.

Difficulties were also experienced at the small-scale steam injection step. This was overcome by use of a redesigned steam injection unit which permitted continuous coagulation without plugging.

Once having injected steam into the juice (see Fig. 2), the curd may either be floated off the top of a separation tank or allowed to settle prior to separation from the brown juice. The separated curd is then dried by drum drier or in a hot air drier holding the product temperature down to 80° or preferably below 60°. Under these conditions xanthophyll losses are held to a minimum.

As our laboratory and pilot plant work progressed, our industry collaborator moved ahead, installing full-scale equipment approximately 50 times as large as ours. We have worked with him in his plant to solve problems as they arose and at one time the bulk of our pilot plant was side by side with his to try to locate the source of problems.

While there is still work to be done to increase throughput and product quality, and to develop markets, the first several carloads of new product (X-PRO brand of PRO-XAN) have been produced and we feel confident that at least several main objectives have been accomplished. While continuing to work out details of the Phase I research (PRO-XAN process), we shall now turn more of our attention to the next phase (II), that of converting the PRO-XAN to human grade LPC plus a stabilized xanthophyll for poultry.

## Editorial note

Differences in objective explain many of the differences in approach in the three preceding papers. Hollo & Koch and Kohler & Bickoff wish primarily to separate the crop into fodder for ruminant and non-ruminant animals. Pirie regards people as the most important of the non-ruminants and so attaches more importance to achieving the maximum practical extraction. Human food is much more valuable than animal fodder.

There are also a few differences that call for more research. Kohler & Bickoff stabilize xanthophyll by coagulating slightly alkaline extracts. This is in accordance with general experience. They also get improved protein texture by alkaline coagulation. This conflicts with observation in other laboratories. We have little experience with lucerne at Rothamsted, but extracts from other species that yield neutral or alkaline saps (e.g. cucumber, marrow, sugar-beet) give a curd that is difficult to handle unless the extract is taken to about pH 5. Alkaline extracts have been made from many other species so as to increase the extractability of the protein, we have had to make these slightly acid to get satisfactory coagulation.

The precise conditions of coagulation have other effects. Leaf extracts from different species are proteolytic to varying extents. This leads to some loss of coagulable protein if there is delay before coagulation. Experiments were reported from New Zealand showing that with lucerne the loss is only 6% after 2 hr at room temperature. Leaf extracts also contain ribonuclease. Degradation of ribosomes by it (Pirie, 1950, 1957; Singh, 1960) may be responsible for part of the apparent protein loss because the TCA precipitate made from fresh extracts will contain more ribonucleic acid than a precipitate made after ribonuclease has had time to act. In certain circumstances, nucleic acid in the diet is detrimental, some autolysis could then be beneficial.

# 8

# Factors Influencing the Availability of Lysine in Leaf Protein

R. M. Allison

Chemical determination of the amino acid composition of a foodstuff seldom leads to a reliable assessment of its biological value (BV) and true digestibility (TD). In particular the application of Carpenter's method (1960) for determining available lysine, while successful in assessing the nutritional quality of feeds compounded of animal protein, was quite unsuccessful initially when applied to mixed animal–plant protein feedstuffs (Carpenter, 1960; Rao *et al,* 1963; Roach *et al,* 1967; Ostrowski *et al,* 1970). However, several modifications of the Carpenter method have been reported recently claiming more or less success (Roach *et al,* 1967; Ostrowski *et al,* 1970).

The increasing awareness of the superior amino acid composition of bulk leaf protein, particularly in respect of the problem amino acids, lysine, methionine, threonine and tyrosine, has prompted the large-scale production of bulk leaf protein from many plant species. These protein concentrates have been biologically tested in small-animal and human feeding experiments. In several cases neither BV or TD have been as great as would be expected from the chemically measured amino acid composition.

During protein extraction there are inevitably reactions between native protein and substances of the polyphenol class, as well as with other compounds. The available lysine was measured in leaf protein concentrates, damaged to controlled extents, and the results correlated with those obtained from nutritional trials. For the chemical determinations a new procedure using deamination with nitrous acid was used. The development of this, with experimental details of the work on leaf protein concentrate, summarized here, is to be published elsewhere (R.M. Allison, W.M. Laird and R.L.M. Synge).

## Methods

Most of the samples were made at Rothamsted several years ago (Morrison & Pirie, 1961). They were crude concentrates containing lipid and other water-insoluble constituents. No attempt was made to exclude air from the preparations either during preparation or storage.

Tea leaf protein (D1, *PSE*(15–23)*PB*), bean leaf protein (A1, *PSE*(18–21) *P*1), and Lucerne I were prepared by the methods of Jennings *et al* (1968); the significance of the identification code is explained in that paper. Oxidation was prevented either by working under $N_2$ or by adding $SO_2$ (cf. Lough, 1968). Lucerne II was prepared similarly but with no attempt to prevent oxidation. Lucerne III was similar to Lucerne II but 26 mg of chlorogenic acid was added to 25 g of fresh leaf tissue at the beginning of the oxidation.

**Nutritional evaluation**

The nutritional results for BV and TD are from Henry & Ford (1965) while those for Protein Efficiency Ratio (PER) and Gross Protein Value (GPV) are from Woodham (1965). The latter author also provided the available-lysine data (Carpenter, 1960).

**Chemical analyses**

*Total N* was determined by semimicro Kjeldahl. *Quantitative amino acid* analyses were performed on a Beckman Spinco Model 120C Amino Acid Analyser. Hydrolysates were prepared by refluxing approximately 5 mg of protein, or the deamination product, for 18 hr with redistilled constant boiling hydrochloric acid. To separate ornithine from lysine, both of which are eluted together on the standard 5 cm column for the analysis of the basic amino acids of hydrolysates, an 18 cm column eluted at pH 5·25 and 55° was used.

Deamination of whole protein was performed under the conditions of Peters & Van Slyke (1932), allowing the reaction to proceed for 30 min at room temperature and scaling down the amounts of reagents and working up as described by Allison, Laird & Synge (to be published).

## Results and discussion

The results of the application of the method to the lucerne leaf protein prepared according to Jennings *et al* (1968), to preparations resulting from deliberate modifications of the method, and to LPC preparations tested by Woodham (1965) are shown in Table 1. The agreement between the figures in the last two columns was sufficiently encouraging for the method to be applied to a wider range of crude LPCs which had been tested in rat feeding

TABLE 1. Lysine analysis and nutritive value

| | 1 | 2 | 3 | 4 | 5 | 6 | 7 | 8 |
|---|---|---|---|---|---|---|---|---|
| Sample | N % of DM | Hydrolysate (Lysine N % of total N) | Hydrolysate of deaminated sample (Lysine N as % of total N before deamination) | Undeaminated Lysine as % of total Lysine | Gross protein value GPV | Protein efficiency ratio PER | Available Lysine (N % of total N) | Available Lysine N % of total N calculated from 2 and 3 |
| | | | | | [All three from Woodham (1965)] | | | |
| Lucerne I | 13·9 | 8·7 | 0·6 | 6·5 | — | — | — | |
| Lucerne II | 14·2 | 7·4 | 0·7 | 9·5 | — | — | — | |
| Lucerne III | 14·2 | 8·0 | 1·5 | 18·5 | — | — | — | |
| Rye | 9·2 | 8·4 | 1·29 | 15·4 | 75 | 2·06 | 6·05 | 7·11 |
| Wheat | 11·2 | 7·1 | 0·93 | 13·1 | 85 | 2·81 | 6·35 | 6·17 |
| Pea | 9·5 | 7·5 | 1·64 | 20·5 | 68 | 1·90 | 6·30 | 5·86 |

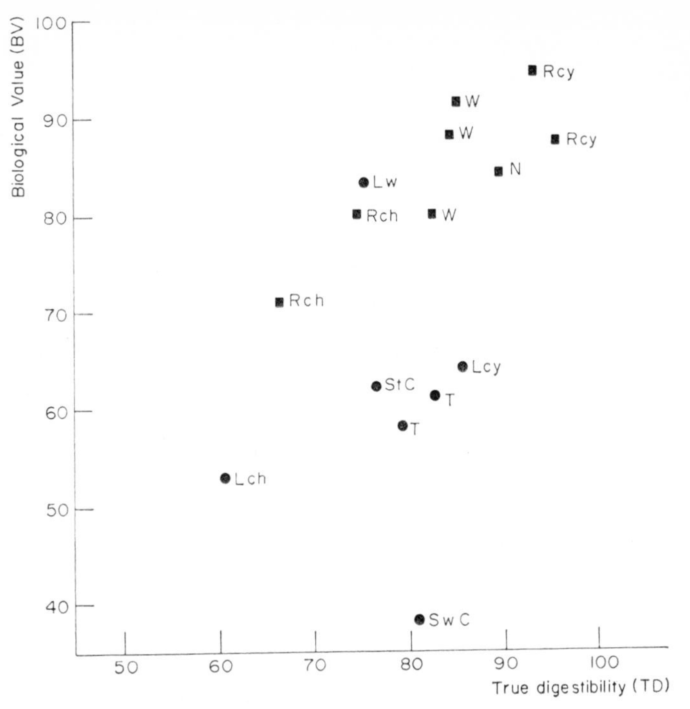


**Figure 8.1. True digestibility.**
**(see page 85 for the coding)**

experiments by Henry & Ford (1965) for different nutritional parameters, i.e. biological value (BV) and true digestibility (TD). The plot of BV versus TD is shown in Fig. 1. The product of these two nutritional parameters (BV $\times$ TD) was taken as a nutritional value figure (NV). Undeaminated lysine as % of the total lysine was then plotted against NV with the result shown in Fig. 2. The relationship between the total lysine and lysine not available for deamination by the nitrous acid procedure is shown in Fig. 3. The correlation is highly significant and the regression line is shown in the figure.

Noteworthy features appear to be the effect on the integrity of the protein of (Lu 1) of careful maintenance of reducing conditions, and the effects of air and added chlorogenic acid on the proportion of undeaminated lysine. It is obvious also that the chloroplastic protein preparations contain much undeaminated lysine. Compared with these, the cytoplasmic protein from

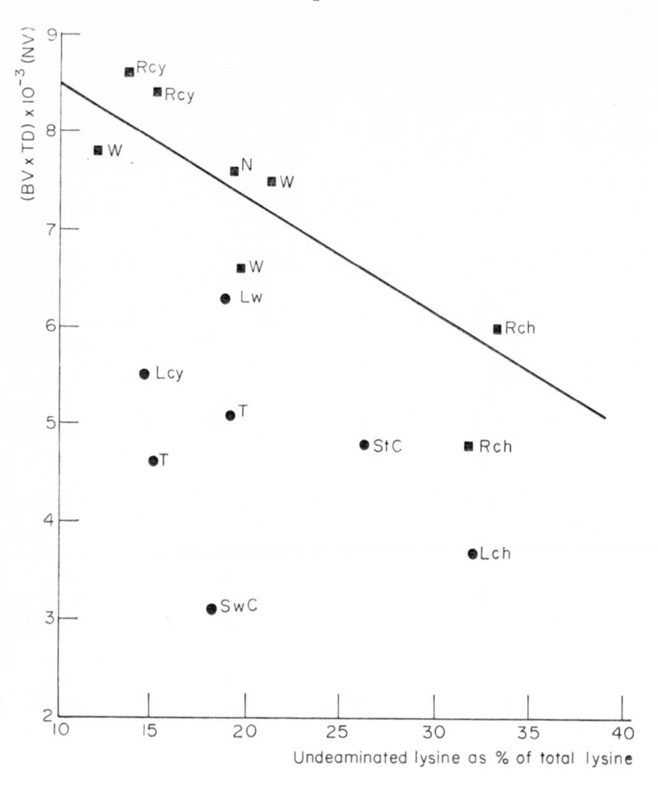


Figure 8.2. Undeaminated lysine as percentage of total lysine.

lupin and rape leaves are less altered both chemically and nutritionally (Fig. 2).

The correlation analysis of all the results for which BV and TD were available, separated into legumes and non-legumes, omitting sweet clover, is shown in Tables 2 and 3. These results, especially those for the non-legumes, help to explain the effect of oxidation in damaging the nutritional value of many LPC preparations. The equivocal data for the legumes suggests that reactive groups other than the ε-amino-group of lysine are important, or that extraneous factors damaging to good nutrition are present.

The oxidation of *o*-diphenols to *o*-quinones by the ubiquitous polyphenolases present in most plant tissues is well established. The topic is reviewed by Pierpoint (1971). Leaf protein is rich in lysine; this is its merit compared

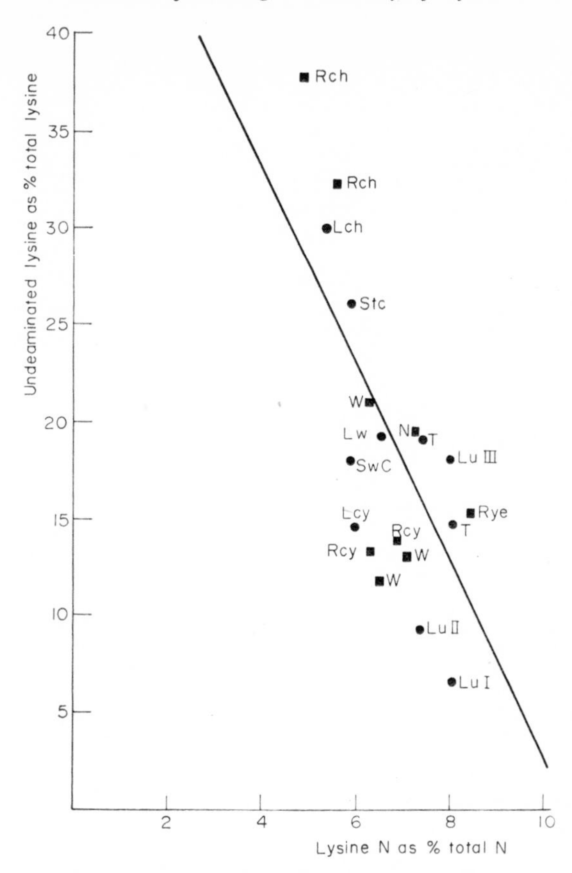


**Figure 8.3.** Lysine N as percentage of total N.

to plant storage proteins. The action of polyphenols and quinones on the ε-amino-groups of this lysine and the subsequent polymerization of the polyphenols into tannin–protein complexes could render large blocks of amino acids, including other essential amino acids, inaccessible to the digestive processes of monogastric animals. Hence a small amount of oxidation could cause a large diminution in nutritional value. This phenomenon has been studied in some detail by Horigome & Kandatsu (1968), who demonstrated the effects of *o*-diphenols and polyphenolases from red clover upon the BV and TD of casein.

TABLE 2. Lysine analysis and nutritive value

| | 1 | 2 | 3 | 4 | 5 | 6 | 7 |
|---|---|---|---|---|---|---|---|
| Sample | N % of DM | Hydrolysate (Lysine N % of total N) | Hydrolysate of deaminated sample (Lysine N as % of total N before deamination) | Undeaminated Lysine as % of total Lysine | Biological value (BV)* | True digestibility (TD)* | BV × TD |
| Tares | 10·93 | 8·1 | 1·21 | 15·0 | 58 | 80 | 4640 |
| Tares | 10·93 | 7·4 | 1·42 | 19·2 | 61 | 83 | 5063 |
| Nasturtium | 10·61 | 7·2 | 1·38 | 19·2 | 84 | 91 | 7644 |
| Sweet clover | 10·59 | 5·9 | 1·07 | 18·2 | 38 | 81 | 3078 |
| Wheat | 11·76 | 5·9 | 1·16 | 19·7 | 80 | 83 | 6640 |
| Wheat | 10·35 | 6·3 | 1·34 | 21·2 | 88 | 85 | 7480 |
| Wheat | 10·38 | 6·4 | 0·77 | 12·0 | 91 | 86 | 7826 |
| Lupin (whole) | 12·00 | 6·6 | 1·28 | 19·4 | 83 | 76 | 6308 |
| Lupin (chloro) | 9·28 | 5·4 | 1·63 | 30·2 | 35 | 61 | 3233 |
| Lupin (cyto) | 13·00 | 6·0 | 0·88 | 14·7 | 64 | 86 | 5504 |
| Strawberry clover | 9·22 | 5·9 | 1·55 | 26·2 | 62 | 77 | 4774 |
| Rape (chloro) | 8·58 | 5·5 | 1·79 | 32·6 | 71 | 67 | 4757 |
| Rape (chloro) | 8·37 | 4·8 | 1·83 | 38·2 | 80 | 75 | 6000 |
| Rape (cyto) | 12·55 | 6·8 | 1·01 | 14·9 | 87 | 96 | 8352 |
| Rape (cyto) | 13·13 | 6·2 | 0·82 | 13·3 | 91 | 94 | 8554 |
| Bean (Al, *PSE* (18–21) *Pl*) | 11·8 | 8·9 | 1·44 | 16·2 | — | — | — |
| Tea (Dl, *PSE* (15–23) *PB*) | 12·4 | 6·6 | 1·39 | 21·0 | — | — | — |

* Henry & Ford (1965).

TABLE 3

| Correlations (data in Table 2) | Legumes | Non-legumes |
|---|---|---|
| Biological value and true digestibility | 0·269 N.S. | 0·829* |
| Undeaminated lysine and nutritional value | – 0·653 N.S. | – 0·856** |
| Lysine nitrogen and undeaminated lysine | – 0·722* | – 0·708* |
| Undeaminated lysine and biological value | – 0·317 N.S. | – 0·784* |

N.S., not significant; *, **, significant at $P \times 0{\cdot}05$ and 0·01 respectively.

---

*Coding for Figures* 1, 2 *and* 3

Lch =Lupin chloroplast protein
Lcy =Lupin cytoplasmic protein
Lu I =Lucerne protein preparation according to Jennings *et al* (1968)
Lu II =Prepared as for Lu I except for omission of $SO_2$
Lu III =Prepared as Lu II with addition of chlorogenic acid
Lw =Lupin whole protein
N =Nasturtium protein
Rch =Rape chloroplast protein
Rcy =Rape cytoplasmic protein
StC =Strawberry clover protein
SwC =Sweet clover protein
T =Tares protein
W =Wheat protein

■, Non-legumes; ●, Legumes.

# 9

# Drying, Preservation, Solvent Extraction and Separation into 'Chloroplast' and 'Cytoplasmic' Fractions

N. W. Pirie

Protein that is allowed to dry slowly in air at room temperature or in an oven at $<80°$ is not nutritionally impaired but it becomes hard, dark and difficult to use. Freeze-drying yields a smooth pale product, especially if the press-cake is rewetted to 20% DM or more before freezing and if the freezing is completed quickly. A convenient unit in which 5–10 kg lots of protein can be frozen in 1–2 min has been described (Pirie, 1964). By adding an extender such as salt (Pirie, 1959a) or meal (Duckworth *et al,* 1961), smooth material can be made by direct drying. Salt can, if need be, be washed out of the protein before it is used, and a carbohydrate extender is not an objectionable component of material that is to be used as human food. There are, however, advantages in making a smooth and protein-rich product by a less expensive method than freeze-drying. Arkcoll (1969) managed this by allowing the protein to dry in air until it contained only 20–30% water, and then grinding it finely during the final stages of drying.

Preservation is sometimes the sole reason for drying. That objective can be more economically and conveniently achieved by salting, canning or pickling. By incorporating acetic acid, at a final concentration of 2%, along with 0·2% of orange-peel oil, into the moist press cake, Subba Rao *et al* (1967) preserved leaf protein for more than a year at room temperature in Mysore.

Chlorophyll, and most of the other material in the leaf that is soluble in lipid solvents, is in the chloroplasts and so is concentrated in the 'chloroplast' fraction. On first acquaintance with unfractionated leaf protein, many people regard the chlorophyll, or its breakdown products, with disfavour. Experience suggests that hostility to a green food is not deeply rooted. Many dark coloured, or even dark green, foods are already eaten and familiarity makes acceptable many appearances and textures that were not at first pleasing.

Nevertheless, the advantages and disadvantages of retaining these non-protein components deserve, and get, attention, and methods by which they can be removed, or kept from reacting with the protein have been investigated.

Leaf lipids are highly unsaturated. Lima *et al* (1965) and Buchanan (1969a) found that more than half the lipid in protein from several species was doubly or trebly unsaturated. These lipids do not react readily with protein in the absence of water, moist protein reacted with lipid in a few hours at 100° (Buchanan, 1969a, b) with loss of digestibility by enzymes and of nutritional value. This reaction can be reversed by solvent extraction and it does not depend on the presence of air. The protein, even in absence of lipid, is, however, slowly destroyed on more prolonged heating in air. These observations help to clarify the effects of drying leaf protein. There is no damage when water is removed at a low temperature (Duckworth *et al*, 1961) and presumably there would not be any if it were removed sufficiently rapidly at a high temperature. It is slow high-temperature drying that is harmful. By taking sufficient care over the conditions in which lipid and water were removed, Buchanan (1969a) made a preparation from wheat leaves with TD=91 and PER=3·2.

Chlorophyll and its breakdown products are easily extracted from heat-coagulated protein by solvent extraction—especially if moist protein is used soon after filtering it off. Work in New Zealand by Lohrey & Chapman (unpublished) shows that pigment removal is facilitated by heat-coagulating the extract after adding a little surfactant (e.g. Tween 20) to it.

It is possible that it will be necessary to remove lipids to get a stable and acceptable product. There are many reasons for hoping that it will not be. Solvent extraction is expensive and if it were made standard, protein production would become an aspect of technologically sophisticated large-scale industry instead of being a simple process that could be managed locally. Furthermore, the lipids are nutritionally valuable. Particular attention has been given to β-carotene (p. 138).

Before coagulation, a leaf extract can be separated into what are loosely called the 'chloroplast' and 'cytoplasmic' fractions. Rouelle (p. 166) noticed the association of the green colour with the coagulum that separated at 50–55°. This remains the standard method of fractionation—it has even been patented. There is a similar separation when extracts are centrifuged at high speed, or precipitated by salts or water-miscible solvents. With some species there is little 'chloroplast' protein in extracts from frozen leaves, and in some this fraction coagulates if the extract is frozen. The precise composition of the fractions probably varies with the method of fractionation and with the physiological state of the leaf, but chlorophyll is always associated with the fraction that separates most easily.

The 'chloroplast' fraction, because it coagulates first, brings out with it most of the dust that preliminary washing may have failed to remove from the leaf surfaces; this may account for the general observation (e.g. Subba Rau *et al*, 1969) that it contains more ash than the 'cytoplasmic' fraction. The point has not been closely examined. It also contains more fibre. Part of this is the chloroplast membrane but part will be leaf fibre incompletely removed by whatever process of straining or centrifuging the extract has been subjected to before coagulation. The quality of a 'chloroplast' fraction is therefore a useful index of the technical skill of those making the original extract. Quinones, polyphenols and other tanning agents are more abundant in the chloroplasts than the cytoplasm of many species and may be expected (Pirie, 1961, 1966a) to combine with protein especially in the presence of air. To varying extents this combination impairs its nutritive value. An important facet of research on leaf protein will be concerned with the selection of species, the management of their husbandry, and improvement in the methods used to separate the protein, so that these combinations can be minimized. (See also Goldstein & Swain (1965).)

'Cytoplasmic' protein, because of the removal of the various components of the extract mentioned above, contains more N than 'chloroplast' fractions. But it is rarely possible to make a bulk preparation containing more than 14% N. Individual enzymes prepared from leaf extracts contain, as would be expected, 16–18% N. The non-protein contaminants of bulk preparations have not been thoroughly investigated: polysaccharides and phenolic compounds are probably predominant. Work with plant virus preparations shows that, by working quickly, excluding air, or adding reducing agents, these forms of contamination can be partly avoided. Unsystematized observations suggest that this is also so with bulk preparations of leaf protein.

Lexander *et al* (1970) coagulated the 'chloroplast' protein from extracts from frozen leaves of ten species heated for 20 min at 53° and at different pHs between 4·5 and 6·0. The 'cytoplasmic' protein was then coagulated at 80°. Figure 1 shows the percentage of the total extracted protein that was in the 'cytoplasmic' fraction. The obvious importance of the pH at which 'chloroplast' protein is heat-coagulated should be borne in mind in all future work on bulk fractionation. The extent to which the ratio of 'chloroplast' to 'cytoplasmic' protein varied with the nutritional state of the leaf is

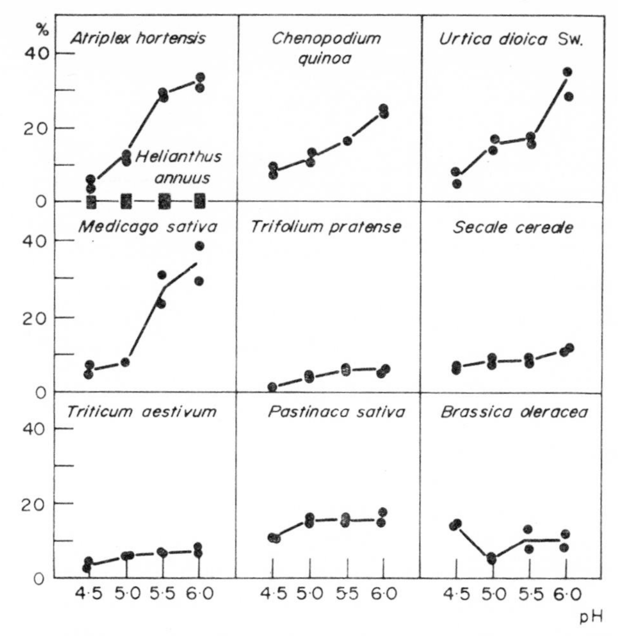


Figure 9.1. Cytoplasmic protein as a percentage of totally precipitated protein when precipitation was induced at different pHs.

not recorded in that paper, but there was a definite increase in the N contens of both fractions as the N content of the original leaf increased. This it presumably because the amount of material contaminating the protein remains nearly constant whereas, with increasing N in the leaf, the amount of protein extracted increases and so is correspondingly less contaminated. The consensus among those making leaf protein can be summed up 'The better the crop the better the product'. Lexander *et al* (1970) record one 'cytoplasmic' preparation with 15·6% N.

Though traditional, heating is not altogether satisfactory as a method of fractionating leaf extracts. Byers (1971) examined the fractions separated from barley, lupin and Chinese cabbage by centrifuging in steps up to 50,000 g for an hour. Her main concern was with the amino acid composition of the various fractions (p. 107), but this work also showed that the percentage of N was greater in the more slowly sedimenting fractions, that age diminished the amount of 'chloroplast' protein extracted but had little influence on the amount of 'cytoplasmic' protein, and that more of the extracted protein was in the 'cytoplasmic' fraction when the separation was made centrifugally, than when it was made by heating—even at such a low temperature as 43°.

Differences in the amino acid composition of the two fractions, though small, appear to be real (p. 109). Differences in nutritional value are greater and are probably caused by the greater tendency of 'chloroplast' protein to form complexes that make the protein less digestible, and some amino acids, e.g. lysine, partly unavailable. Lexander *et al* (1970) and Byers (1971) measured digestibility by enzymes *in vitro*. 'Cytoplasmic' protein was 80–90% digestible, that is to say, converted into material soluble in TCA, whereas 40% was a more usual value with 'chloroplast' protein. In accordance with the generalization suggested above, the greater the N content of the original leaf, the more digestible the 'chloroplast' fraction made from it. Considerable differences have been noted between species. Detailed discussion of these would be premature until many more preparations, from each species, have been studied. It is probable that the differences are caused in part by differences in the extent to which complexes have formed, and that will depend on the time elapsing at each stage of the preparation, on the amount of aeration, and on the temperatures at which the extracts were held. Many of the preparations studied were not made with these points in mind. It is therefore possible that many of the differences that appear in the literature on the digestibility and nutritive value of preparations from different species are not intrinsic differences between the proteins but are the result of uncontrolled reactions during the purification.

Measurements of the digestibility of an insoluble protein by enzymes *in vitro* are not easy to interpret because the surface exposed to enzyme attack depends on the fineness of grinding. There is no evidence that undenatured leaf protein is less readily attacked than other proteins, but all comparisons concur that dried material is less readily or less completely digested by enzymes such as papain, pepsin and trypsin than proteins such as casein. It is likely that combination with lipids, quinones or other tanning agents is responsible for the impaired digestibility. Such an effect is shown in experiments with tanned proteins (Hawkins, 1959; Horigome & Kandatsu, 1968; Feeny, 1969). Experiments *in vitro* do not resemble at all closely the turbulent conditions in an animal's gut; there is therefore little reason to be surprised that digestion *in vivo* tends to be better than experiments *in vitro* might have suggested. In spite of their insolubility, protein preparations from several leaf species supported the multiplication of *Tetrahymena* better than did casein, though not so well as egg protein. There were, however, species differences; protein from lucerne was the least suitable for *Tetrahymena* (Lexander *et al*, 1970).

# Section III
# Composition and Nutritional Value

# 10

# The Amino Acid Composition of Some Leaf Protein Preparations

M. Byers

The introduction of automated systems for the chemical analysis of amino acids during the last decade has led to a vast increase in the literature on the composition of proteins. A new technique does not automatically invalidate results obtained by older methods, though it may increase the existing difficulty of comparing results because analytical methods differ in precision. In protein analysis there is a further complication because many different methods of hydrolysis can be used, some of which adversely affect the determination of one or more amino acids. Confusion is increased by the several ways in which results can be expressed, and also by the many publications which give no indication of the accuracy of the determination (e.g. the amount, or percentage, of N calculated from the recovered amino acids). Results, however obtained, should always be treated with a certain reserve if this vital piece of information is missing.

Because of these strictures it is possibly useful to discuss the three stages in determining the amino acid composition of a protein (hydrolysis of the sample; analysis of the amino acids; and presentation of results), with particular reference to the analysis of leaf proteins, before reviewing the literature. It should perhaps be stressed here that this paper is concerned solely with the composition of protein(s) extracted from leaves, and not the amino acid composition of the whole leaf, sometimes confusingly referred to as 'leaf protein'.

## Hydrolysis of the sample

All leaf proteins, except those precipitated from chloroplast-free extracts, contain appreciable amounts of lipid, and some ash and carbohydrate. The problems associated with the hydrolysis of impure proteins were discussed by Lugg (1946), and much attention has since been given to the hydrolysis of samples, such as animal feeds and seed proteins, which contain large amounts of fats and/or carbohydrates (Dustin *et al*, 1953; Smith *et al*, 1965; Roach, 1966). Reviews by Tristram (1966) and Roach & Gehrke (1970) on the same topic emphasize the need for time and temperature studies, because of the different rates of both release and destruction of certain amino acids. Ideally three hydrolyses should be done on every sample, each for a different time, and the results for all amino acids plotted and extrapolated to zero hours. However, unless one has unlimited time, or a large analytical capacity, one analysis per sample must be the norm for many, and it is difficult initially to decide on what conditions to use. It is probably best to study the literature and find those most often used for similar samples, try these, and if successful stick to them. No absolute rules can be laid down for the hydrolysis of a protein, as conditions will be governed by the nature of the sample and, what is sometimes overlooked, the equipment available—a point often relevant in developing countries. However, the following general points should be observed: assuming the sample is homogeneous, and is finely ground, hydrolyse a small quantity in a sealed tube instead of refluxing a larger amount: use a large volume of acid, e.g. 500 times the weight of sample taken (this diminishes humin formation and loss of certain amino acids): keep the temperature constant throughout the period of hydrolysis: use the same temperature for all hydrolyses: hydrolyse in an inert atmosphere, either under $N_2$ or *in vacuo* (Crestfield *et al*, 1963; Boulter, 1966; Eaker, 1968) (preferably the latter as air dissolved in the acid can also be removed—see below).

The following conditions have been used successfully at Rothamsted for the hydrolysis of leaf proteins (Byers, 1971).

1. Air-equilibrate all samples before use (proteins absorb about 10% moisture).

2. Weigh a 4–10 mg sample into a Pyrex tube (180 × 18 mm): take samples for N (micro-Kjeldahl) and dry matter (100° for 24 hr) determinations at the same time.

3. Add 6N (re-distilled) or constant boiling HCl to the protein: 1·0 ml for each 2·0 mg sample.

4. Freeze tube and contents (in acetone–dry ice mixture), evacuate and allow to thaw slowly so as to remove the air dissolved in the HCl as well, seal *in vacuo* (0·5 mm Hg or less), and keep at 110° for 18 hr.

5. Take hydrolysate to dryness under reduced pressure in a rotary evaporator (first filtering through a No. 3 sintered glass filter stick to remove humin

when present), free from excess HCl and redissolved in 25 ml 0·01N HCl containing nor-leucine (0·1 mM). (Removing the HCl by desiccating over NaOH is not recommended as loss of several amino acids can occur, notably threonine, serine and glutamic acid, which can react with each other or form equilibrium states after partial conversion to another compound (Ikawa & Snell, 1961; Crestfield *et al*, 1963; Smyth & Elliott, 1964; Eaker, 1968).

For column chromatography an internal standard must be included in the mixture to be analysed: this is invariably nor-leucine for protein hydrolysates. It can be added to the sample before hydrolysis, after hydrolysis but before removing the HCl, or after removing the HCl. At which stage it is added is largely a matter of individual choice: there are arguments for and against at each stage. For analysis by other methods the hydrolysate should be redissolved in an appropriate medium.

Ideally a hydrolysate should be processed immediately incubation is finished: it can be stored at –20° in the sealed tube but once opened processing should be as quick as possible (Eaker, 1968). The redissolved hydrolysate can be kept for short periods at –20°, but is best used immediately (Eaker, 1968).

Of the eighteen amino acids commonly found in proteins, seventeen can be determined on an acid hydrolysate: tryptophan is destroyed and must be determined separately. Using the technique described above satisfactory results can be obtained for all amino acids except cystine, which is partially destroyed, especially in those samples containing much non-protein material. The measurement of methionine, as the sum of methionine and its sulphoxides, is reliable. Both these amino acids can be quantitatively determined after conversion to cysteic acid and methionine sulphone respectively by a carefully controlled oxidation of the protein by performic acid, usually at –10°, before hydrolysis (Bidmead & Ley, 1958), but because the other amino acids are adversely affected two separate analyses are required, and the results have to be combined. When absolute values of these two amino acids are required performic oxidation before hydrolysis is probably the best method to use, but judging by the wide scatter of the reported results for cystine in leaf proteins obtained by oxidation methods, from 0·5 to 2·4% (i.e. g amino acid per 100 g recovered amino acids), there seems some doubt about the reliability of the method with this sort of sample. Tryptophan determinations are usually based on the specific colour reaction between it and *p*-dimethylaminobenzaldehyde (Spies & Chambers, 1948) after hydrolysis of the protein by alkali (either sodium or barium hydroxide) (Spies & Chambers, 1949; Miller, 1967) or the bacterial proteolytic enzyme Pronase (Spies, 1967).

### Analysis of the amino acids

The methods for the analysis of a mixture of amino acids can be grouped as follows:

1. Two-dimensional paper chromatography of the free acids or certain derivatives, e.g. dinitrophenyl (DNP) amino acids.
2. Column chromatography of the free acids on ion exchange resins.
3. Gas chromatography of a volatile acetylated amino acid ester.
4. Microbiological assay of the free acids.

Of these, gas chromatography is relatively new and is not, as yet, widely used. Most of the reported results have been on simple systems and as far as is known it has not been applied to hydrolysates of complex proteins.

Separation by two-dimensional chromatography, introduced around 1945, was the universally used chemical method until superseded by ion exchange techniques in the early 1960s. The free amino acids separated on paper chromatography were developed with ninhydrin, and the intensity of the resultant spots either measured directly or photometrically after elution of the colour from the cut out spots. DNP-amino acids, which are visible as yellow spots without further development, are measured in a similar way. Paper chromatography is still used, particularly when sophisticated equipment is unavailable: however, there are dangers in relying solely on results from this method. Separation of leucine and iso-leucine is seldom complete and a combined result has to be given: the less abundant amino acids can easily be overlooked, particularly when spots are adjacent: and finally, the uncertainty of the method is about 10%. Paper chromatography must now be regarded as a semi-quantitative technique only, from which approximate values can be estimated: its use is thoroughly unacceptable when looking for small differences in composition in a series of samples. The amino acid composition of leaf protein is well established, and no more approximate observations are needed, therefore this method is obsolete as far as they are concerned: further the small differences that have been found between the various preparations would not be detectable.

Most analyses are now done by column chromatography on ion exchange resins, and the amino acids in a hydrolysate can be resolved in anything from 2 to 21 hr according to the system used. The method, briefly, consists of a buffer gradient passing, under pressure, over a cationic resin on which a sample of the hydrolysate has been adsorbed. Both the rate and order of release of the amino acids depends on the nature of the buffer, and they are affected by pH, ionic concentration, and additives such as alcohol. The continuous stream of eluant is mixed, usually in an inert atmosphere, with a solution of ninhydrin and heated: when an amino acid is present the reaction mixture develops the characteristic purple (or yellow) colour given by ninhydrin–amino acid complexes. The intensity of these bands is measured photometrically at 570 nm (440 nm for the yellow complexes) and recorded on a moving chart. From the peak areas obtained the amount of each amino acid can be calculated by reference to the area on a standardizing chromatogram made by a known amount of the specific amino acid. The accuracy of this method depends a great deal on the operator. Because the quantities analysed are small (from 0·05 to 0·15 μmole of each amino acid is the usual range), the analytical solutions, samples and apparatus must be carefully prepared to avoid contamination. Given attention, reproducibility within 5% should be consistently obtained (and with further care narrowed to 3%). The percentage accuracy of an analysis is arrived at by calculating the N found in the recovered amino acids and expressing it as a percentage of the known N content of the sample.

Microbiological assay of total amino acids was introduced some 25 years ago, and at the time was the most sensitive method available. It was found that if certain micro-organisms, particularly *Lactobacilli*, were grown in a medium containing all the nutrients required for growth except the amino acid one wanted to assay, addition of this acid, in small quantities, stimulated the growth of the test organism in proportion to the amount added. Total amino acid composition is found by using an acid hydrolysate of the protein: recently this technique has been successfully applied to estimating 'available' amino acids in a protein using an enzymic hydrolysate (Ford, 1962, 1964).

### Presentation of results

It is disconcerting on entering this field to find how many ways are used to express results: μmole per ml of hydrolysate; μmole %; g residue or g amino acid per 100 g sample; g residue or g amino acid per 100 g recovered residues or recovered amino acids; g amino acid per 16 g N; g amino acid N per 16 g N and amino acid N as % of total recovered N or protein N; this list is far from complete. This state of affairs does not arise entirely through capriciousness on the part of researchers, certain methods of expression are better for one aspect of protein work than another, e.g. residues are best used when establishing amino acid sequence, but it causes confusion when collaborative work between groups with different backgrounds is required. Certain styles, though useful in further calculations, should not be published: μmole per ml of hydrolysate means nothing, except to the operator, because it is not related to the weight or N content of the sample taken: μmole % can perhaps be used to compare one set of results with another, but should never be presented as though it were an absolute value. One suspects laziness on the operator's part when results are published in this form. Nutritionists, who use g amino acid per 16 g N, are probably the most consistent group. In leaf protein publications four styles of presentation are commonly used: to illustrate the differences between these (and one other way) the amino acid composition of an unfractionated wheat leaf protein, made at Rothamsted, is given in Table 1. Results, as g amino acid per 100 g sample, or leaf protein (N.B. never 'protein', which is not the same thing), must be calculated before g amino acid per 100 g recovered amino acids can be arrived at, but should *never* be published because one preparation cannot be compared with another, as samples have different protein contents. A useful check for operators is to calculate g residue per 100 g sample because its total should approximate to the value for the protein content of the preparation—see Table I.

## Comparison of analyses on a single sample

It is not easy, especially in the absence of recovery figures, to decide what constitutes a 'good' analysis. The methods used for hydrolysis and determination of the amino acids, as well as the experience and reliability of the operator, all have to be considered. There have been from time to time collaborative tests on amino acid determination, the most recent was that organized by the Protein Quality Group of the UK Agricultural Research Council (Porter *et al*, 1968). Here, the amino acids in a standard mixture and in the hydrolysates of cod muscle and gelatin were measured by ion exchange chromatography in nine laboratories. Good agreement was obtained using the standard mixture, but more varied results were obtained for the hydro-

TABLE 1. The amino acid composition (tryptophan excluded) of an unfractionated wheat leaf protein expressed in five different ways: N as % of DM on sample=9·60: protein content (N×6·0)=57·6

| Amino acid | g residue per 100 g sample (or leaf protein) | g amino acid per 100 g sample (or leaf protein) | g amino acid per 100 g recovered amino acids | g amino acid per 16 g N | g amino acid N as % of total recovered N |
|---|---|---|---|---|---|
| Aspartic acid | 5·41 | 6·25 | 9·70 | 10·48 | 6·89 |
| Threonine | 2·78 | 3·27 | 5·06 | 5·48 | 4·03 |
| Serine | 2·53 | 3·05 | 4·72 | 5·11 | 4·26 |
| Glutamic acid | 6·37 | 7·26 | 11·25 | 12·17 | 7·24 |
| Proline | 2·69 | 3·19 | 4·93 | 5·34 | 4·06 |
| Glycine | 2·77 | 3·64 | 5·64 | 6·11 | 7·11 |
| Alanine | 3·56 | 4·46 | 6 91 | 7·47 | 7·34 |
| Valine | 3·47 | 4·10 | 6·34 | 6·86 | 5·13 |
| Cystine | 0·84 | 0·98 | 1·52 | 1·65 | 1·20 |
| Methionine | 1·30 | 1·48 | 2·29 | 2·47 | 1·45 |
| Isoleucine | 2·74 | 3·17 | 4·91 | 5·31 | 3·54 |
| Leucine | 5·08 | 5·89 | 9·13 | 9·87 | 6·59 |
| Tyrosine | 2·55 | 2·83 | 4·38 | 4·74 | 2·29 |
| Phenylalanine | 3·50 | 3·92 | 6·08 | 6·57 | 3·48 |
| Ammonia | 0·91 | 0·91 | 1·41 | 1·52 | 7·83 |
| Lysine | 3·77 | 4·29 | 6·65 | 7·20 | 8·61 |
| Histidine | 1·33 | 1·51 | 2·34 | 2·53 | 4·28 |
| Arginine | 3·91 | 4·36 | 6·75 | 7·30 | 14·67 |
| | 55·51 | 64·56 | 100·01 | 108·18 | 100·00 |

lysates of the two proteins. In part this can be attributed to the several methods of hydrolysis used, and the authors themselves point out that further trials are desirable to establish the best procedures for measuring cystine and methionine.

A similar collaborative trial using an unfractionated wheat leaf protein made at Rothamsted has recently been started. Of the six laboratories involved, four have so far returned results: these are given in Table 2, along with the conditions used for hydrolysis and analysis. The results from laboratories C and E, where substantially the same conditions for hydrolysis were used, are in reasonably good agreement, as are the results from D, except for the S-containing amino acids which would have been affected by air. Of the results A and B, obtained by the same laboratory on different hydrolysates and on different instruments, A was obtained from a new instrument by a new operator, and is to be repeated.

While no new series of analyses are required on unfractionated leaf proteins, more collaborative tests on samples originating from different laboratories would be useful, both to the analysts and the maker of the protein.

### Amino acid composition of unfractionated leaf protein

Chibnall (1939) noted a similarity in composition between the various protein fractions extracted from the leaf, and between preparations made from leaves of different species. This general similarity was confirmed by later workers, but at the same time some differences in composition were noted, especially in the contents of the basic and S-containing amino acids. According to Smith & Agiza (1951) and Steward *et al* (1954) the composition of leaf protein was influenced by leaf age, and, using grass and lupin respectively, they showed that protein extracted from mature leaves contained less of the basic amino acids than protein made from young leaves. A decreased methionine content, with a corresponding increase in cyst(e)ine was also associated with increasing leaf age (Lugg & Weller, 1948).

These and many other early results were gathered together in a review compiled by Kuppuswamy *et al* (1958) which was immensely useful at that time, though the distinction between extracted protein and whole leaf preparations was not always entirely clear. Since then more accurate methods have been introduced, and most of these old results, particularly those for the more labile amino acids, do not compare well with those obtained during the last decade.

TABLE 2. The amino acid composition of an unfractionated wheat leaf protein (tryptophan excluded) obtained by five analyses from four laboratories (results expressed as g amino acid per 100 g recovered amino acids)

| Laboratory | A | B | C | D | E |
|---|---|---|---|---|---|
| Wt of sample (mg) | 25 | 25 | 9 | 500 | Not stated |
| Vol. of 6N or constant boiling HCl (ml) | 125 | 125 | 4·5 | 150 | but this laboratory always uses 1 : 500 sample to HCl (same as laboratory C) |
| Other additives | 15 mg $SnCl_2$ | 15 mg $SnCl_2$ | Nil | Nil | Nil |
| Hydrolysis conditions | Reflux for 24 hr | Reflux for 24 hr | 110° for 18 h, sealed tube *in vacuo* | 110° for 24 h, reflux in air | 110° for 20 h, sealed tube *in vacuo* |
| Instrument used | E.E.L. 2-column | Technicon, 2-column | Technicon, single-column | E.E.L. 2-column | Beckman, 2-column |
| AMINO ACID | | | | | |
| Aspartic acid | 9·50 | 9·26 | 9·70 | 9·90 | 9·57 |
| Threonine | 5·22 | 4·92 | 5·06 | 4·85 | 5·25 |
| Serine | 4·56 | 4·17 | 4·72 | 4·37 | 4·87 |
| Glutamic acid | 13·91 | 12·70 | 11·25 | 11·84 | 11·22 |
| Proline | 2·16 | 4·85 | 4·93 | 4·56 | 5·10 |
| Glycine | 5·96 | 5·90 | 5·64 | 5·83 | 5·54 |
| Alanine | 7·01 | 7·25 | 6·91 | 6·99 | 6·57 |
| Valine | 5·98 | 6·45 | 6·34 | 6·31 | 6·35 |
| Cystine | 0·78[1] | 0·77[1] | 1·52 | 0·00 | 1·06[1] |
| Methionine | 2·08[12] (averaged) | 1·19[1,2] (averaged) | 2·29[2] | 2·14[2] | 2·46[2] |
| Isoleucine | 5·07 | 5·36 | 4·91 | 5·05 | 5·08 |
| Leucine | 9·45 | 9·24 | 9·13 | 9·42 | 9·22 |
| Tyrosine | 4·78 | 4·68 | 4·38 | 4·47 | 4·31 |
| Phenylalanine | 6·50 | 6·40 | 6·08 | 6·21 | 6·40 |
| Ammonia | 3·46 | — | 1·41 | 1·65 | 0·79 |
| Lysine | 5·77 | 6·45 | 6·65 | 6·50 | 6·30 |
| Histidine | 2·22 | 2·46 | 2·34 | 2·62 | 2·79 |
| Arginine | 5·61 | 7·25 | 6·75 | 7·28 | 7·13 |
| | 100·02 | 100·02 | 100·01 | 99·99 | 100·01 |

[1] Cystine or methionine determined after oxidation to cysteic or methionine sulphone.
[2] Methionine determined directly on normal hydrolysate.

Many of Chibnall's original preparations have now been analysed by ion exchange chromatography, and their similarity in composition confirmed (Chibnall *et al*, 1963). On the basis of these and other recent results, it is generally agreed that the amino acid composition of unfractionated, or whole, leaf protein from different species is similar, and that it is not affected by the physiological age or state of the plant, or by fertilizer treatment (Bryant & Fowden, 1959; Pleshkov & Fowden, 1959; Chibnall *et al*, 1963; Gerloff *et al*, 1965; Wilson & Tilley, 1965; Byers, 1971). Nevertheless, variation in content is greater for some amino acids than for others, and Gerloff *et al* (1965) in giving the range for some thirty preparations made from eight species showed that, S-containing amino acids apart, the largest differences were in the amounts of proline and lysine, which ranged from 3·5 to 5·4 and 4·5 to 7·2% (i.e. g amino acid per 100 g recovered amino acids) respectively. Less importance should be attached to the results for proline, because of the greater inaccuracy of this determination, but the differences in lysine content were real. Chibnall *et al* (1963) thought that the amount of total lysine in acid hydrolysates might be underestimated by 10–15%, and recently Byers (1971) found that proteins contained from 10 to 15% less lysine when coagulated by heat then when precipitated by acid from the same extract. Small but consistent differences have also recently been found between the contents of aspartic acid, alanine and methionine in a series of barley, lupin and Chinese cabbage preparations, indicating that these amino acids might be species dependant (Byers, 1971).

The S-containing amino acids always present problems because of their instability. The published methionine contents of leaf proteins are seldom directly comparable, as they depend both on the method of analysis, and more particularly on the conditions in which the sample is hydrolysed. Omitting those results where some destruction of methionine would occur during hydrolysis, protein extracted from cereal and grass leaves (i.e. Gramineae) seems to contain slightly more methionine than protein from other species (Chibnall *et al*, 1963; Gerloff *et al*, 1965; Byers, 1971). It is also difficult to compare the published results for cyst(e)ine for similar reasons. Theoretically a controlled oxidation should give a quantitative conversion to cysteic acid, which can then be measured chromatographically, but the scattered results obtained by oxidation methods, already referred to, differ little from results obtained by the direct determination of cystine: this emphasizes the doubts about the reliability of oxidation methods when applied to leaf proteins. Because of the nutritional importance of the total-S amino acid content some other method for determining cyst(e)ine is needed, if only as a check on results obtained by more usual procedures. It is possible that measuring the total-S content of leaf protein might give a reliable estimate of the combined methonine+cyst(e)ine content (Miller & Naismith, 1959), the cyst(e)ine could then be estimated by difference. A total-S determination has been applied to various foodstuffs and diets (Miller & Donoso, 1963; Pellett *et al,* 1969), and its relative lack of success has been attributed to the lysine-deficient state of the product being examined. This method, however, might be applicable to leaf protein, as it contains an adequate amount of lysine and should be free from non-protein S which also confuses this estimation.

Most published amino acid analyses are on proteins extracted from leaves grown in a temperate climate, and the papers by Chibnall *et al* (1963), Gerloff *et al* (1965), Wilson & Tilley (1965) and Byers (1971) present results on preparations made from a large number of species and from leaves in different physiological states. Some analyses on lucerne (alfalfa) proteins taken from these papers (and one unpublished result) can be compared in Table 3 (N.B. results are presented as g amino acid per 100 g recovered amino acids, which has necessitated converting some of the published figures). There are fewer published results on protein extracted from tropical leaves but analyses done at Rothamsted (Byers, 1970b) on a variety of preparations made at CFTRI, Mysore show that their composition resembles that of other leaf proteins. The methionine content of proteins extracted from eighteen species of leaves grown in India (Valli Devi *et al,* 1965), though a little less than in the proteins from CFTRI, was of the same order (some was probably destroyed during hydrolysis) but the amount of lysine reported was only half, or less, what is now regarded as usual. It is unfortunate that these latter figures feature so prominently in the latest edition of 'Table of the amino acids in food and foodstuffs' (Harvey, 1970), anyone unaware of other analyses might assume that all leaf proteins are lysine-deficient. The other references given are obscure, and also misleading: none of the publications I cite in this article are quoted. Preparations made from twelve species of leaves grown in Ceylon have been analysed for total amino acid content (Sentheshanmuganathan & Durand, 1969), but the results are badly expressed and cannot be compared one with another without further calculation: the percentage recoveries are also poor. However, the results given by Eggum (1970b) on protein extracted from three Nigerian-grown species, including cassava (*Manihot utilissima*) leaves, are nearer to the usual values. The rather

TABLE 3. The amino acid composition of some unfractionated lucerne (alfalfa) (*Medicago sativa*) preparations determined on different samples by various analysts (results expressed as g amino acid per 100 g recovered amino acids)

| Analyst | Gerloff *et al* (1965) | Chibnall *et al* (1963) | Wilson & Tilley (1965) | Smith (1966) | Byers (1970b) | Byers (unpublished) |
|---|---|---|---|---|---|---|
| Sample origin | British Glues & Chemicals | Chibnall | Wilson & Tilley (1965) | British Glues & Chemicals | CFTRI, Mysore | Rothamsted |
| Aspartic | 10·2 | 9·43 | 9·77 | 9·88 | 9·70 | 10·56 |
| Threonine | 5·1 | 5·73 | 4·64 | 5·09 | 4·99 | 5·12 |
| Serine | 4·3 | 5·18 | 4·09 | 4·58 | 4·22 | 4·69 |
| Glutamic acid | 11·40 | 11·69 | 10·64 | 10·79 | 11·55 | 11·38 |
| Proline | 4·8 | 4·76 | 4·87 | 4·58 | 4·55 | 4·81 |
| Glycine | 5·7 | 5·41 | 5·26 | 4·79 | 5·40 | 5·65 |
| Alanine | 6·4 | 6·03 | 5·95 | 5·91 | 6·27 | 6·29 |
| Valine | 6·3 | 6·26 | 6·52 | 5·91 | 6·45 | 6·37 |
| Cystine | 0·6[1] | 1·54[1] | 1·07[1] | 0·81 | 1·54[2] | 0·63[2] |
| Methionine | 1·9[3] | 2·22[3] | 1·66[4] | 1·83 | 2·05[3] | 1·97[3] |
| Isoleucine | 6·6 | 5·02 | 5·40 | 5·40 | 5·21 | 5·29 |
| Leucine | 9·6 | 9·58 | 9·20 | 9·78 | 9·43 | 9·57 |
| Tyrosine | 4·5 | 4·57 | 4·84 | 5·09 | 4·73 | 4·46 |
| Phenylalanine | 6·4 | 4·13 | 6·06 | 6·21 | 6·11 | 6·14 |
| Ammonia | — | 0·90 | 1·81 | 1·43 | 1·75 | 1·45 |
| Lysine | 6·3 | 6·90 | 6·43 | 6·42 | 6·83 | 6·85 |
| Histidine | 2·1 | 2·31 | 2·47 | 2·04 | 2·45 | 2·33 |
| Arginine | 5·8 | 6·18 | 6·78 | 7·43 | 6·77 | 6·43 |
| Tryptophan | 1·90 | 2·17 | 2·54 | 2·04 | — | — |
| | 99·9 | 100·01 | 100·00 | 100·01 | 100·00 | 99·99 |

[1] Cystine determined after oxidation to cysteic acid.
[2] Some destruction of cystine during hydrolysis.
[3] Accurate methionine determination.
[4] Some destruction of methionine during hydrolysis.

primitive coagulation technique used (boiling in an open vessel for 5–10 min) probably accounts for the decreased values of some amino acids, especially lysine.

## Amino acid composition of fractionated leaf proteins

The distribution and composition of the fractions which comprise whole leaf protein have, until recently, received less attention. Most protein in the leaf is in the chloroplasts, and the remainder is divided between the other organelles and the 'soluble' fraction. The term 'chloroplastic' protein is widely, and loosely, used to denote the fraction containing most of the chlorophyll, but its exact nature will depend on the method of separation used. Thus, a fraction made by controlled heating of an extract will contain cell debris and co-precipitated material, and possibly some protein not associated with the chlorophyll-containing fraction. A more precise separation can be achieved by centrifuging an extract at increasing speeds. The protein precipitable from an extract after removing the chlorophyll-containing fraction is usually referred to as the 'cytoplasmic' protein; its composition depends on the method used to separate the chloroplastic material.

The ratio of chloroplastic and cytoplasmic fractions made by the controlled heating of leaf extracts has been measured with several species (Byers & Davys, 1964; Byers, 1965; Subba Rau *et al*, 1969): chloroplastic fractions have also been made by flocculating the leaf extract with calcium chloride (Yemm & Folkes, 1953) as well as by centrifuging the whole extract (Chayen *et al,* 1961; Wilson & Tilley, 1965), and cytoplasmic protein by acidifying (Yemm & Folkes, 1953) or adding alcohol (Wilson & Tilley, 1965) to the chloroplast-free extracts. A recent systematic study of barley, lupin and Chinese cabbage (*Brassica chinensis*) extracts (Byers, 1971) made from leaves of different ages has shown that the chlorophyll-containing fraction sedimented less readily in extracts from older leaves, though the ease of sedimenting varied with species. When this fraction is removed, the amount of protein-N remaining in the extract is the same, irrespective of leaf age, though, again, the amount differed with species.

Early reports on the amino acid composition of chloroplastic and cytoplasmic fractions were conflicting, and comparison of the results was difficult because most of them were based on the analysis of a single preparation. It was, however, generally agreed that their composition resembled that of unfractionated protein, though this observation was often qualified by

TABLE 4. The amino acid composition of some chloroplastic, unfractionated and cytoplasmic proteins from the leaves of barley, lupin and Chinese cabbage (*Brassica chinensis*): (expressed as g amino acid per 100 g recovered amino acids)

| | Mean percentages of amino acids from three preparations in each group (excluding cystine, ammonia and tryptophan) | | | | | | | | |
|---|---|---|---|---|---|---|---|---|---|
| | Chloroplastic | | | Unfractionated | | | Cytoplasmic | | |
| Amino acid | Barley | Lupin | Chinese cabbage | Barley | Lupin | Chinese cabbage | Barley | Lupin | Chinese cabbage |
| Aspartic acid | 9·75 | 10·10 | 9·86 | 9·57 | 10·22 | 10·01 | 9·62 | 10·01 | 10·05 |
| Threonine | 4·82 | 4·97 | 4·87 | 5·07 | 5·01 | 5·22 | 5·41 | 5·04 | 5·43 |
| Serine | 4·85 | 5·15 | 5·24 | 4·40 | 4·68 | 4·50 | 4·10 | 4·13 | 4·12 |
| Glutamic acid | 11·00 | 11·35 | 11·28 | 11·41 | 11·88 | 11·91 | 11·94 | 12·15 | 12·21 |
| Proline | 4·88 | 5·06 | 4·98 | 4·68 | 4·79 | 4·72 | 4·62 | 4·79 | 4·46 |
| Glycine | 6·12 | 5·97 | 6·53 | 5·64 | 5·69 | 5·35 | 5·38 | 5·32 | 5·29 |
| Alanine | 7·05 | 6·40 | 6·45 | 6·71 | 6·21 | 6·10 | 6·52 | 5·99 | 5·91 |
| Valine | 6·16 | 6·10 | 5·64 | 6·37 | 6·27 | 6·06 | 6·50 | 6·32 | 6·17 |
| Methionine | 2·28 | 1·90 | 2·11 | 2·24 | 1·70 | 1·94 | 2·39 | 1·76 | 2·13 |
| Isoleucine | 5·25 | 5·76 | 5·03 | 4·95 | 4·93 | 4·62 | 4·74 | 4·42 | 4·38 |
| Leucine | 10·43 | 10·68 | 10·40 | 9·33 | 9·75 | 9·29 | 8·42 | 9·21 | 8·79 |
| Tyrosine | 4·49 | 4·20 | 4·19 | 4·50 | 4·61 | 4·71 | 4·92 | 5·56 | 5·07 |
| Phenylalanine | 6·97 | 7·16 | 6·85 | 6·22 | 6·24 | 6·24 | 5·84 | 5·82 | 5·87 |
| Lysine | 5·60 | 4·78 | 5·23 | 6·61 | 6·60 | 7·08 | 7·06 | 7·30 | 7·23 |
| Histidine | 1·82 | 1·91 | 2·00 | 2·34 | 2·31 | 2·42 | 2·66 | 2·82 | 2·65 |
| Arginine | 6·29 | 6·13 | 6·33 | 6·89 | 6·35 | 6·46 | 7·01 | 6·67 | 6·89 |

excepting one, or more, amino acids (not always the same one). According to Yemm & Folkes (1953) a cytoplasmic fraction prepared from barley leaf extract contained more lysine than the corresponding unfractionated protein. Increased amounts of both lysine and histidine were found by Wilson & Tilley (1965) in a cytoplasmic fraction from lucerne. Also citing a lucerne cytoplasmic fraction, Smith (1966) reported that it contained more histidine and less leucine than the unfractionated protein, but the lysine content of the two preparations was similar. It has now been established (Byers, 1971), using preparations made from barley, lupin and Chinese cabbage extracts, that chloroplastic and cytoplasmic proteins do not have the same composition: both contain similar amounts of aspartic acid, proline, alanine valine and methionine, and differ only slightly in their contents of threonine, serine, glutamic acid, isoleucine, tyrosine, phenylalanine and arginine, but there is less leucine and substantially more histidine and lysine in the cytoplasmic than in the chloroplastic fraction. As with unfractionated protein, the composition of these fractions is not influenced by leaf age. These differences in composition between the various fractions, and between species (referred to in the previous section), show most clearly when the results are classified statistically: the averaged results for the three types of preparation from the three species examined are shown in Table 4.

### Comparison with the FAO reference protein

The ranges for the nutritionally essential amino acid content of some unfractionated leaf proteins (as g amino acid per 100 g recovered amino acids)

TABLE 5. The range of amino acid analyses (expressed as g amino acid per 100 g recovered amino acids) on fifty-six unfractionated leaf proteins made from twenty-one species

| Amino acid | | FAO (1965) provisional recommendation |
|---|---|---|
| Isoleucine | 4·5– 5·5 | 4·2 |
| Leucine | 8·8–10·2 | 4·8 |
| Lysine | 5·6– 7·3 | 4·2 |
| Methionine | 1·6– 2·6 | 2·2 |
| Phenylalanine | 5·5– 6·8 | 2.8 |
| Threonine | 4·7– 5·8 | 2·8 |
| Tryptophan | 1·2– 2·3 | 1·4 |
| Tyrosine | 3·7– 4·9 | 2·8 |
| Valine | 5·9– 6·9 | 4·2 |
| Cystine values uncertain | | |

are shown in Table 5, along with the provisional pattern for the FAO (1965) reference protein. In contrast to seed proteins these preparations are rich in lysine, and the average content is more than one and a half times that of the FAO reference protein. Although the amount of tryptophan is occasionally below the recommended level, in terms of absolute amount, methionine is the first limiting essential amino acid.

The methionine content of chloroplastic and cytoplasmic fractions, which resembles that of the corresponding unfractionated protein, is also limiting in these preparations, despite the variation in the amounts of leucine, histidine and lysine.

## Amino acid composition in relation to nutritive value

It is necessary to determine the total amount of each amino acid in a protein to establish its overall composition, and to find the limiting essential amino acid. Nevertheless, in the absence of other factors, it is usually the availability of the essential amino acids, rather than their absolute amounts, which determine the nutritive value of a protein. Lysine, because of the many reactions which render it unavailable, and methionine, which is often limiting, are especially important in this context. The small differences in amino acid composition between the various leaf protein fractions are unlikely to account for the proved nutritional superiority of cytoplasmic over both unfractionated and chloroplastic protein, and other reasons must be sought.

In terms of absolute amount methionine is the limiting essential amino acid in all leaf protein preparations, and its addition improves diets containing unfractionated protein (Henry & Ford, 1965; Shurpalekar *et al*, 1969; Gordon & Topps, 1970). Adding lysine (Subba Rau *et al*, 1969; Shurpalekar *et al*, 1969) or isoleucine (Henry & Ford, 1965) to the same diet has no effect, and when either is combined with methionine the total response is no more than with methionine alone (Henry & Ford, 1965; Shurpalekar *et al*, 1969). Care should be taken, however, in translating these results to humans, as these experiments were done on rats, which have a greater methionine requirement than most other species (Eggum, 1970a). As the total methionine content is sometimes marginal relative to the FAO reference protein, the poorer *in vivo* performance of unfractionated leaf proteins (compared to whole egg or casein) has been attributed to methionine deficiency (Henry & Ford, 1965; Shurpalekar *et al*, 1969). However, as the nutritive value of cytoplasmic protein, with a similar methionine content, approaches that of casein (Henry & Ford, 1965), it would seem that the quantities of essential amino acids present are reasonably balanced with respect to one another. This, coupled with the evidence on methionine supplementation (Henry & Ford, 1965; Shurpalekar *et al,* 1969; Gordon & Topps, 1970), indicates lack of availability rather than absolute amount in unfractionated protein. From the even poorer *in vivo* results with chloroplastic protein it seems that this unavailability might be associated with the presence of chloroplastic material. This now seems probable: 'available' methionine (determined micro-biologically using *Streptococcus zymogenes*) (Ford, 1964) in some barley proteins ranged from 100% in a cytoplasmic fraction to around 40% in a sedimented chloroplastic fraction (Ford, 1970). About 90% was 'available' in unfractionated leaf proteins (Ford, 1964, 1970).

Because of the interdependence of dietary methionine, cyst(e)ine and some of the S-containing compounds which have a sparing action on methionine (Miller & Samuel, 1968), it is probably more accurate to take the total S-containing amino acids into account when discussing protein values, rather than the methionine alone. Hence the earlier emphasis given to finding a reliable way of determining cyst(e)ine.

The lack of extra response when lysine is added to diets containing unfractionated protein (Subba Rau *et al,* 1969; Shurpalekar *et al,* 1969) indicates that enough is nutritionally 'available', and measurements, using Carpenter's (1960) method, show that from 70 to 80% of the total content is in this form (Woodham, 1965; Glencross, 1969). Recently van Slyke's nitrous deamination procedure (Peters & van Slyke, 1932) (which removes any ε-amino groups of lysine not bound to quinones and polyphenols) has been used on various preparations (p. 78). The amount of 'available' lysine was measured as the difference between the total lysine content of the intact protein and the amount determined in a deaminated sample of the same protein. Results on some leaf proteins agreed well with those obtained by other chemical methods (Woodham, 1965). The position of lysine in the chloroplastic fractions, both sedimented and heat-precipitated, is somewhat different. The total content is not much above the recommended FAO minimum (4·2%), and if some was unavailable, as is probable, these fractions could be regarded as lysine-deficient. So far as is known supplementing chloroplastic proteins with lysine, or any other amino acid has not been tried, and values for 'available' lysine in these fractions have not been reported.

**Factors affecting the availability of certain amino acids**

There are many possible reasons for the unavailability of the S-containing amino acids in unfractionated leaf protein. Heat-processing is known to depress the availability of methionine in certain food-stuffs without affecting the total amount determined by chemical methods (Miller *et al,* 1965; Ellinger & Boyne, 1965; Woodham & Dawson, 1968), and Ellinger & Palmer (1969) have recently shown that this could be due to the oxidation of peptide-bound methionine to the sulphoxide (metO). Hydrolysates of heat-coagulated proteins usually contain more metO than hydrolysates of other types of preparations (Byers, 1970a), but it is not known whether this is produced during the extraction process or during hydrolysis, or if it arises from a combination of both. More metO is also found in hydrolysates of proteins processed slowly than in quickly-made preparations (Byers, 1970a). These aspects need investigating, in all protein fractions, but for self-evident reasons chemical methods alone would be inadequate, and alternative techniques such as rat tests (Ellinger & Palmer, 1969) and/or micro-biological assay (Ford, 1964) would also have to be used.

Because of the almost total lack of hydrolysis with unactivated papain it was thought there were few free thiol groups in unfractionated leaf protein (Byers, 1967). This view was substantiated when it was found that cytoplasmic protein was digested to the same extent as casein (around 80%) by the unactivated enzyme (Byers, unpublished), implying no shortage of free thiol groups in this fraction. The *o*-quinones produced by the enzymic oxidation of chlorogenic and caffeic acids in plant extracts are known to couple with the thiol groups of both free and peptide-bound cysteine (Pierpoint, 1969a, b), and it could be that these reactions are responsible for the apparent unavailability of the –SH in preparations containing chloroplastic material. Lipid protection of thiol groups seems unlikely as only a slight increase in digestibility (both *in vivo* and *in vitro*) follows extraction of unfractionated protein with 2:1 (v/v) chloroform: methanol (Byers, 1967; Buchanan, 1969a, b).

*o*-Quinones and polyphenols also react with the ε-amino group of lysine (Pierpoint, 1969a, b), and could be responsible for loss of availability of this amino acid. Decreased availability, however, is more conventionally associated with the Maillard reaction between the ε-amino group of lysine and aldehyde groups, which occurs on heating many protein-containing foodstuffs. This last reaction probably occurs to some extent during heat-coagulation, for leaf extracts are rich in reducing sugars, but there seems insufficient condensation to cause much loss of nutritive value. The difference in total lysine content between heat- and TCA-precipitated proteins (which has been observed in both unfractionated and cytoplasmic preparations) is not fully explained by the Maillard reaction, though condensation products, resistant to acid hydrolysis, may be formed (Bujard *et al,* 1967).

The poorer nutritive value of unfractionated and chloroplastic proteins compared with the cytoplasmic fraction, probably has more than one cause, and many of the reactions considered above are probably occurring simultaneously. Heat-coagulation, by steam-injection, of unfractionated extracts has been standard at Rothamsted for many years; it is preferred to other methods of precipitation when working on a large scale because the coagulum filters easily, whereas the finer precipitates obtained by pH adjustment or adding alcohol need centrifuging. Heat-damage to lysine, while probably occurring to a small extent, is insufficient to cause much loss of nutritive value. However, the reactions rendering methionine, and possibly cyst(e)ine, unavailable are imperfectly understood, and at present heat-damage cannot be excluded.

## Conclusions

The amino acid composition of the various fractions extractable from the leaf is similar, though less leucine and more histidine and lysine are found in cytoplasmic protein (i.e. that precipitated from chloroplast-free extracts) than in the chloroplastic fractions. These real, but small, differences are, however, unlikely to account for the nutritional superiority of cytoplasmic over both chloroplastic and unfractionated protein. Methionine, the limiting amino acid in all leaf protein preparations, appears to be 'available' in sufficient quantity in cytoplasmic, but not in unfractionated or chloroplastic protein. There is sufficient lysine 'available' in cytoplasmic and unfractionated protein, but it may be marginal in some chloroplastic preparations. Reasons for the unavailability of methionine, cyst(e)ine and lysine are put forward, and several of the reactions considered are probably occurring simultaneously.

It would be economical to use all the protein that is extractable from the leaf, and supplement it, if necessary, with methionine, to improve its nutritive value. But there is an argument for centrifuging extracts from species that contain a large percentage of cytoplasmic protein, especially those from which the chloroplastic fraction sediments easily: e.g. lupin. The colourless cytoplasmic protein does not need supplementing with methionine, and would overcome two objections often raised against the use of leaf protein in human nutrition, the green colour of the unfractionated product, and its smaller nutritive value compared to milk and egg. The chloroplastic fraction need not be wasted: it could be fed to animals. The ratio of chloroplastic to cytoplasmic protein in a leaf extract is constant irrespective of leaf age, but is species dependent. Work has started on extracts from other species to determined both the ratio of the two fractions, and the ease with which the chloroplast fraction sediments.

## 11

# The Use of Animal Tests for the Evaluation of Leaf Protein Concentrates

A. A. WOODHAM

Leafy vegetation has provided the main, and often the only, source of nutrients for ruminant livestock from earliest times until the present day. Possessing a digestive system which permits the break-down of cellulose-containing entities within the leaf, ruminants are enabled to draw upon an energy source which is unavailable to animals with simple stomachs, and also to supplement deficiencies in the protein by virtue of their self-engendered microbial population. In 1942, however, Chibnall showed that in all probability the nutritive value of the protein in leaves required no boosting with microbes, for he was able to extract from cocksfoot grass a material which contained all of the amino acids known to be essential for the rat, and in just the right amounts (Pirie, 1942). The realization that leaves were capable of yielding a concentrate suitable for meeting the needs of the more discriminating non-ruminant stimulated interest in processes for producing leaf protein in commercial quantities. Nutritionists in western pig- and poultry-producing countries were well aware of the restrictions imposed by the necessity of including expensive materials such as fish meal and soyabean meal as sources of good quality protein, and the possibility of a cheap alternative protein source from leaves was highly attractive. Additionally, of course, there was the pressing need for a high quality protein source to supplement the diets of protein-starved peoples in many parts of the world. Clearly nutritional experiments with animals were necessary to test the correctness of the deductions drawn from amino acid analysis, and also to ensure that there were no harmful factors present of sufficient importance to rule out the use of LPC in the feeding of man and animals.

### Nutritional studies using rats and poultry, prior to 1956

Preliminary studies on fractions isolated from cocksfoot grass were carried out by Davies *et al* (1952) and the results were disappointing. Digestibility was low compared with conventional feedingstuffs. The average digestibility for forty common materials listed by Block & Mitchell (1946) was 92% while values found by Davies *et al* for 'cytoplasmic' and 'chloroplast' protein fractions respectively were 61% and 75%. A protein extracted from timothy gave a digestibility as low as 40%. Biological values determined by the Mitchell-Thomas method with rats were also low, but again 'cytoplasmic' protein was superior to 'chloroplast' protein.

Intensive programmes of research initiated at the Grassland Research Institute, Hurley, Berks., and the Rowett Research Institute, Aberdeen, resulted in a series of papers published between 1952 and 1956 (Carpenter *et al*, 1952, 1954; Ellinger, 1954; Cowlishaw *et al*, 1956). These studies were summarized and examined critically by Duckworth & Woodham (1961). At both centres the biological criterion used was a chick growth measurement of the supplementary value of the sample when fed as part of a cereal based diet adequately provided with minerals and vitamins. This test—the Gross Protein Value (GPV) test—was developed by Heiman *et al* (1939) and later adapted for use with British diets (Carpenter *et al*, 1952; Duckworth *et al*, 1961). Of some twenty-seven leaf protein concentrates produced by a variety of processes from clover, lucerne and grass (chiefly Italian ryegrass) only two or three had GPVs approaching that which would have been expected from the amino acid composition. An estimate of Net Protein Utilization (NPU) was made by Bender on a sample of LPC prepared by one particular commercial process and this yielded an extremely low value indeed (Bender, 1956).

Using laying birds, cereal diets supplemented with perennial ryegrass LPC were compared with fish meal-supplemented diets (Hughes & Eyles, 1953a, b, c). There was no difference in performance so far as egg number and egg weight were concerned, but a tendency for the LPC diets to produce greenish albumins led the authors to suggest that extraction of chlorophyll might be necessary to avoid the possibility of greenish yolks forming during storage. Poorer growth and egg laying performance were obtained when lucerne LPC was used (Cowlishaw & Eyles, 1956).

The generally poor quality indicated in these experiments may be attributed to several causes. The low digestibility demonstrated by Davies *et al* (1952) is an inherent characteristic of samples produced by the methods used and is doubtless attributable to their relatively large content of chloroplastic protein. Unavailability of lysine due to the processing conditions is also a possibility, particularly in view of the fact that lysine supplementation produced an improvement in the nutritive value for chicks (Raymond & Tilley, 1956). The choice of leaf was not ideal. Though possessing good amino acid composition (Cooney *et al*, 1948), lucerne was found to contain a saponin-like growth depressant, the effect of which was diminished, though not removed entirely, by including cholesterol in the diet (Cowlishaw *et al*, 1956; Peterson, 1950). It was shown subsequently, too, that grasses produced concentrates greatly inferior to some other leaves such as those of cereals.

That the fibrous residue left after extraction of the bulk of the protein was a suitable ruminant feed was shown in digestibility trials with sheep (Raymond & Harris, 1957). These workers found, however, that silage made from the residue was unpalatable for all classes of stock, and recommended that the material should be fed directly as roughage. This observation appears surprising in view of the satisfactory results achieved in carefully controlled digestibility trials using Romney sheep carried out recently in New Zealand (p. 136). Workers in the Applied Biochemistry and Grassland Divisions of D.S.I.R., Lincoln, found that preparations of either vacuum-pack silage or dried or fresh leaf protein-residue gave satisfactory digestibilities as follows:

Dried residue—restricted intake: 62·6%
Dried residue—*ad libitum* feeding: 63·6%
Wet residue—*ad libitum* feeding: 72·8%
Silage—restricted intake: 69·5%

## Nutritional studies with rats and poultry since 1956

After the early series of experiments described above had been concluded effort was directed towards investigating the factors which were considered to be responsible for the unsatisfactory results. Various leaf species were studied including cereals, rape, mustard, kale, pea vine, potato haulm, etc., and alternative methods of processing were explored.

### Studies of processing conditions

That the avoidance of heating during drying could result in a product of high nutritive value for pigs was demonstrated by Barber *et al* (1959). Using a mixture of material prepared from barley, grass, oats, wheat, rye and kale, thawed out from the frozen state immediately before incorporating into diets, these workers found no difference in growth or efficiency of feed utilization between fish meal and LPC when each provided equivalent amounts of protein.

In a comparison of LPC samples produced from mixed grasses, barley, kale, rye and tares, it was found that the grass protein concentrate was inferior to the others as judged by Gross Protein Value, while the remaining four products all produced chick growth equivalent to that given by soyabean meal (Duckworth & Woodham, 1961). All of the samples were produced by a procedure which involved acetone extraction (Table 1). Next, barley LPC was dried in five different ways—by laboratory-scale freeze drying, by

TABLE 1. GPV of acetone extracted LPC

| Source | GPV |
|---|---|
| Mixed grasses | 71 |
| Barley | 77 |
| Kale | 75 |
| Rye | 82 |
| Tares | 82 |

absorbing the LPC moisture on the basal cereal portion of diets which had been previously dried to below 2% moisture, by acetone drying with a final heating to 30° *in vacuo*, by drying at 40° a 1:1 mixture of LPC and barley, and finally by air-drying at 100° (Table 2). The final treatment was the only one which involved the use of high temperature and this was the only one

TABLE 2. Effect of method of drying barley LPC on nutritive value

| Drying method | GPV |
|---|---|
| 1. Moisture absorbed by dried cereal | 83 |
| 2. Freeze drying (laboratory) | 87 |
| 3. Acetone drying with final heating to 30°C *in vacuo* | 80 |
| 4. 1:1 LPC/barley meal mixture, dried at 40°C | 80 |
| 5. Air dried at 100°C | 39 |

which gave a seriously reduced GPV. The other four samples all fell within the range of GPV given by good quality soyabean meals (80–87). Drying temperature was investigated using rye LPC. This was dried in a forced draught oven at various temperatures and for periods of varying length, the conclusion being that damage to the nutritive value occurred only when the temperature of the material was greater than 82°. Various samples heated at temperatures ranging from 42° to 82° had GPVs of 81–85, but when the temperature was as high as 94° the GPV of the product was only 65. Finally a comparison was made between LPC dried by the contact plate-vacuum dehydration procedure, dried skimmed milk powder, cottonseed meal and groundnut meal in a rat growth test of the Protein Efficiency Ratio type in which each of the concentrates provided 5% of protein along with 5% of cereal protein and 1·4% of yeast protein (Duckworth & Woodham, 1961). Under these conditions LPC was inferior to dried skimmed milk powder, similar to cottonseed meal and superior to groundnut meal (Table 3). From this experiment it was concluded that the product of the commercial drying

TABLE 3. Protein Efficiency Ratios of diets containing various sources of supplementary protein after 2 or 4 weeks on the diets

| | Protein Efficiency Ratio* | | | |
|---|---|---|---|---|
| | 2 weeks on diet | | 4 weeks on diet | |
| | Males | Females | Males | Females |
| Dried skimmed milk | 2·70±0·06 | 2·27±0·06 | 2·68±0·06 | 2·17±0·06 |
| Leaf protein concentrate | 2·16±0·06 | 1·99±0·07 | 2·26±0·05 | 2·06±0·06 |
| Cottonseed meal | 1·96±0·06 | 1·90±0·07 | 1·97±0·05 | 1·94±0·06 |
| Groundnut meal | 1·68±0·10 | 1·83±0·08 | 1·75±0·09 | 1·81±0·07 |

* Protein Efficiency Ratio=g gain per g protein eaten.

process already tested with chicks in the previous experiment was of satisfactory nutritive value and compared well with conventional materials.

It would appear from these experiments that there is reasonable latitude in the drying conditions which may safely be applied to LPC and that damage occurs only in the 82–94° zone.

A further trial using the commercially-dried product was carried out with pigs (Duckworth *et al*, 1961). LPC was compared with fish meal (FM) in diets based on barley and millers' offals at various levels of total and supplementary protein (Table 4). With about 7% of LPC (Diet LPC3) the rate of growth and efficiency of feed utilization were as good as for diets containing 8% of FM (Diet FM1), these two diets being similar in their content of total and supplementary protein. Further increasing the LPC level to 10% did not give any better growth, but the efficiency of utilization of the feed was significantly improved. Of particular interest was the satisfactory result achieved by mixing a small proportion of LPC with groundnut meal. This combination produced as good growth and as high an efficiency of utilization of food as did diets containing much higher levels of LPC, and is a good

TABLE 4. The nutritive value of leaf protein concentrate for pigs

| Diet | Protein content of diet (%) | Amount of supplementary protein (%) | Time to reach 100 lb live-weight (days) | Daily gain (lb) | Feed eaten per lb gain (lb) |
|---|---|---|---|---|---|
| Fish meal 1 | 16·1 | 4·9 | 51 | 1·32 | 2·79 |
| Fish meal 2 | 14·8 | 3·4 | 56 | 1·19 | 2·95 |
| Fish meal 3 | 13·5 | 1·8 | 57 | 1·15 | 3·11 |
| Leaf protein concentrate 1 | 18·6 | 8·1 | 53 | 1·26 | 2·53 |
| Leaf protein concentrate 2 | 17·1 | 6·2 | 51 | 1·30 | 2·53 |
| Leaf protein concentrate 3 | 15·5 | 4·3 | 54 | 1·23 | 2·78 |
| Leaf protein concentrate 4 | 14·4 | 2·8 | 61 | 1·08 | 3·06 |
| Leaf protein concentrate/ groundnut meal | 19·0 | 5·9 | 53 | 1·24 | 2·59 |
| Standard errors of means | | | ±0·05 | ±0·07 | |
| Standard errors of differences | | | ±0·08 | ±0·10 | |

example of the ability of LPC to improve diets largely composed of poor quality plant proteins. In this respect it behaves similarly to FM.

The satisfactory nature of products prepared by acetone-drying, freeze-drying, and drying on starch in a current of air was confirmed by the work of Henry & Ford (1965) using the balance sheet method for protein evaluation with young rats. The results of the rat tests agreed well with microbiologically determined available lysine, but not with the figure indicated by the chemical fluorodinitrobenzene method (Ford, 1964). With higher concentrations of papain, however, the microbiological method gave higher values, suggesting that the leaf protein was digested slowly.

This work also confirmed the observations of Duckworth & Woodham that oven drying at 100° is harmful, the product of such treatment being much reduced in biological value and in true digestibility. Similar conclusions have been reported by Subba Rau & Singh (1970) using the rat PER method for the evaluation of the heated samples.

Buchanan (1969a) compared the products of various methods of drying in BV, NPU and PER tests with rats, finding that neither moist heat alone, moist heat followed by chloroform extraction, nor extraction with acidified solvents affected BV significantly, but all decreased true digestibility, NPU and PER. He noted that the loss in nutritive value on heating could be reversed by mild solvent extraction.

**Stage of maturity of leaf, and leaf species studies**

Henry & Ford (1965) showed that Biological Value and True Digestibility increased with increasing maturity of the leaf; this observation is contrary to that of Davies *et al* (1952). Because the quantity of extractable protein decreases with stage of maturity, the inference from the results of Henry & Ford is that there is an optimum harvest time. In general, preparations from different species of leaf were of the same order as those of good vegetable protein concentrates, confirming the results reported previously by Duckworth & Woodham (1961). Species differences were, however, apparent in the results of the biological value studies reported by Henry (1963) (Table 5) and also by Woodham (1965) (Table 6). However, while Woodham found mustard LPC to be inferior to cereals and grasses for the chick, Henry found

TABLE 5. Biological value of samples of LPC

| | |
|---|---|
| Barley | 81·4±1·41 |
| Wheat | 80·3±1·08 |
| Rye | 75·8±1·01 |
| Mustard | 73·6±1·08 |
| Pea | 65·2±1·82 |
| Tares | 58·3±1·08 |
| Clover | 53·7±2·53 |

TABLE 6. Biological and chemical evaluation of various leaf protein concentrates

| | 1959 | 1961 | | | 1963 | | |
|---|---|---|---|---|---|---|---|
| | | | PER | | | | |
| Source of concentrate | GPV | GPV | Chick | Rat | GPV | PER Chick | ALV (g/16 g N) |
| Wheat | | 61 | 1·67 | 1·84 | 85 | 2·18 | 5·22 |
| Tares | 82 | 72 | 1·78 | 1·76 | | | |
| Rye | 82 | | | | 75 | 2·06 | 5·04 |
| Barley | 77 | | | | | | |
| Kale | 75 | | | | | | |
| Mixed grasses | 71 | | | | | | |
| Pea | | 51 | 1·49 | 1·83 | 68 | 1·90 | 5·21 |
| Maize | | | | | 64 | | |
| Clover | | 54 | 1·35 | 1·58 | | | |
| Mustard | | 50 | 1·35 | 1·56 | | | |

it equivalent for the rat. Both workers found clover LPC to be of relatively poor quality. It may be, of course, that the differences noted were not due to species difference at all, but to differences in stage of maturity of the leaf. Processing can, however, be ruled out in these experiments, as all samples were dried by methods shown to be innocuous.

Singh used rat growth tests in a study of differences between plant species, and concluded that the differences found were attributable rather to feed intake and digestibility than to protein quality (Singh, 1970). The nutritional variations could not be explained on the basis of the minor quantitative differences in amino acid pattern. He found LPC prepared from the Cruciferae superior to lucerne LPC, while material prepared from carrot, *Tecoma* and *Dolichos lablab* failed to support growth. Horse gram, french beans and groundnuts yielded products which were inferior in rat growth tests to lucerne (Singh, 1968).

More detailed results embodying rat growth tests and metabolic studies were presented by Singh at the IBP Technical Group Meeting in Coimbatore. These confirmed the wide variation in nutritive value between samples of LPC prepared from various species. Lucerne, cabbage, knolkhol and radish all yielded protein-efficiency ratios (PER) ranging from 1·5 to 1·9 while rats lost weight on *Dolichos lablab, Sesbania sp.,* carrot and horse gram LPC. The metabolic studies yielded somewhat conflicting results however, materials having PER ranging from 0 to 0·91 all showing Biological Values (BV) ranging from 70 to 77. Satisfactory biological values are reported for lucerne, knolkhol, turnip and cauliflower LPC by Reddy (unpublished). The range and number of results which are now available appear to remove any remaining doubts regarding the existence of protein quality differences attributable unequivocally to difference in species.

Protein efficiency ratio estimations with rats have also been carried out on LPC fractions from lucerne (Subba Rau *et al*, 1969), confirming the superiority of cytoplasmic protein over chloroplastic. These authors also compared the coagulated leaf protein with whole spray-dried extract of leaves, finding the latter to be unsuitable as food because of the association of various undesirable water-soluble constituents which are removed in the normal procedure for preparing LPC.

### Amino acid supplementation experiments

Though possessing an extremely good amino acid pattern the original claim of Chibnall that LPC can provide all of the amino acids known to be essential for the rat requires some modification in the light of newer knowledge of amino acid requirements, and of the improved techniques for amino acid analysis which are now available. Further, one must bear in mind that LPC would not be used as a complete diet in itself, but inevitably it would be mixed with other protein-containing feedingstuffs and its true value must be assessed in the light of its ability to make good the deficiencies in such materials. Like fish meal and soyabean, LPC might be predicted to be a good source of the lysine needed in rations composed largely of cereals and this has indeed been shown to be the case. LPC shares with other plant proteins, however, the disadvantage that it contains rather too little methionine to render it the universally acceptable cereal supplement. True, methionine is cheap and the desirability of incorporating it must be considered in the light of the economic position as a whole. That LPC-cereal diets are only on the borderline of deficiency for the chick may be seen (Table 7) and the pig's requirements are adequately met (Table 8). However, rats benefit from methionine supplementation when fed on LPC alone (Shurpalekar *et al*, 1969). Under these conditions 15% total protein was found to give optimum results, but even with the added methionine the rats did not grow as well as on skimmed milk powder fed at a corresponding level of protein. However the best results from methionine supplementation were obtained at the 10% protein level. Lysine adequacy in the LPC was demonstrated by the failure to obtain any growth response on inclusion of lysine with LPC (Table 9) (Shurpalekar *et al*, 1969). Reddy found that methionine alone or in combination with leucine and/or lysine, improved growth and PER of weanling rats on a diet which contained lucerne LPC as the sole source of protein. She also showed that lysine was ineffective when added as the sole amino acid supplement (Table 10).

The beneficial effect for rats of adding methionine to LPC diets designed for Ghanaians has been demonstrated by Miller (1965). Supplementation of diets containing 10, 20 and 40% of LPC with 0·1, 0·2 and 0·4% of DL-methionine respectively produced corresponding responses from the animals. Recently Eggum (1970b) has confirmed the advantages of supplementary methionine in rat diets using cassava LPC. Low biological value and net protein utilization for the unsupplemented LPC (44–57 and 30–40 respectively) were raised to 80 and 57 respectively on adding methionine.

*The importance of animal tests.* Biological testing of a new food destined for human consumption is considered mandatory nowadays and even when

TABLE 7. Essential amino acid composition of chick diets (g/100 g)

| Supplement | Leaf protein | Fishmeal | Soyabean meal | Groundnut meal | Meat meal | Requirement; chick |
|---|---|---|---|---|---|---|
| Supp. protein (%) | 12 | 12 | 12 | 12 | 12 | |
| Barley protein (%) | 6 | 6 | 6 | 6 | 6 | |
| Total protein (%) | 18 | 18 | 18 | 18 | 18 | |
| Threonine | 0·84 | 0·72 | 0·72 | 0·54 | 0·58 | 0·65 |
| Glycine | 0·96 | 1·43 | 0·84 | 0·91 | 1·85 | 0·85 |
| Valine | 1·11 | 0·95 | 1·0 | 0·83 | 0·84 | 0·8 |
| Cystine+Methionine | 0·70 | 0·79 | 0·73 | 0·54 | 0·54 | 0·72 |
| Isoleucine | 0·83 | 0·74 | 0·86 | 0·65 | 0·54 | 0·8 |
| Leucine | 1·59 | 1·26 | 1·45 | 1·20 | 1·17 | 1·2 |
| Tyrosine+Phenylalanine | 1·82 | 1·34 | 1·64 | 1·61 | 1·20 | 1·3 |
| Lysine | 1·07 | 1·18 | 1·11 | 0·66 | 0·90 | 1·1 |
| Histidine | 0·42 | 0·39 | 0·46 | 0·40 | 0·38 | 0·35 |
| Arginine | 1·18 | 1·13 | 1·25 | 1·55 | 1·20 | 1·0 |
| Tryptophan | 0·33 | 0·26 | 0·26 | 0·22 | 0·23 | 0·2 |

The total provision of essential amino acids in 18% protein chick diets based on barley with various protein supplements. Values underlined fall short of the requirement figures which have been selected from a variety of sources.

TABLE 8. Essential amino acid composition of pig diets (g/100 g)

| Supplement | Leaf protein | Fishmeal | Soyabean meal | Groundnut meal | Meat meal | |
|---|---|---|---|---|---|---|
| Supp. protein (%) | 6 | 6 | 6 | 6 | 6 | Requirement; |
| Barley protein (%) | 10 | 10 | 10 | 10 | 10 | pig |
| Total protein (%) | 16 | 16 | 16 | 16 | 16 | (9–20 kg) |
| Threonine | 0·68 | 0·62 | 0·62 | 0·53 | 0·55 | 0·44 |
| Valine | 0·95 | 0·87 | 0·91 | 0·82 | 0·82 | 0·44 |
| Cystine+Methionine | 0·71 | 0·77 | 0·73 | 0·63 | 0·63 | 0·61 |
| Isoleucine | 0·69 | 0·64 | 0·69 | 0·60 | 0·54 | 0·61 |
| Leucine | 1·34 | 1·17 | 1·26 | 1·14 | 1·13 | 0·61 |
| Tyrosine+Phenylalanine | 1·53 | 1·30 | 1·44 | 1·44 | 1·23 | 0·44 |
| Lysine | 0·86 | 0·91 | 0·88 | 0·65 | 0·77 | 0·96 |
| Histidine | 0·37 | 0·36 | 0·41 | 0·36 | 0·35 | 0·17 |
| Tryptophan | 0·26 | 0·22 | 0·26 | 0·20 | 0·21 | 0·13 |

The total provision of essential amino acids in 16% protein pig diets based on barley with various protein supplements. Values underlined fall short of the requirement figures which have been selected from a variety of sources.

TABLE 9. Growth responses and protein efficiencies at different protein levels and the supplementary effects of lysine and/or methionine fortification, using 21-day-old weanling rats over a 4-week period

| Expt | Diet | Weight gain (g) | Protein intake (g) | PER |
|---|---|---|---|---|
| 1 | 10% lucerne protein | 25·3 | 18·3 | 1·39 |
| | 10% lucerne protein+MET | 61·4 | 22·2 | 2·77 |
| | 10% lucerne protein+MET+LYS | 53·5 | 22·0 | 2·40 |
| 2 | 10% lucerne protein | 16·9 | 14·2 | 1·16 |
| | 15% lucerne protein | 32·2 | 21·6 | 1·41 |
| | 15% lucerne protein+MET | 63·8 | 28·5 | 2·23 |
| | 15% lucerne protein+MET+LYS | 56·1 | 26·0 | 2·16 |
| 3 | 10% lucerne protein | 17·1 | 13·7 | 1·21 |
| | 20% lucerne protein | 46·5 | 33·3 | 1·39 |
| | 20% lucerne protein+MET | 68·0 | 39·2 | 1·72 |
| | 20% lucerne protein+MET+LYS | 66·2 | 37·7 | 1·75 |

MET=methionine; LYS=lysine.

TABLE 10. Mean data related to weight gain, protein intake and protein efficiency ratio of weanling rats fed Lucerne protein and supplements of three limiting amino acids singly and in different combinations and the control protein casein

| Diet | Weight gain (g) | Food intake (g) | Protein intake (g) | PER |
|---|---|---|---|---|
| Basal diet (10% Lucerne protein) | 21·5 | 133 | 13·3 | 1·61 |
| Basal diet+Leucine | 21·8 | 139 | 13·9 | 1·56 |
| Basal diet+Lysine | 23·8 | 142 | 14·2 | 1·68 |
| Basal diet+Methionine | 44·6 | 171 | 17·1 | 2·60 |
| Basal diet+Leucine+Lysine | 20·1 | 134 | 13·4 | 1·50 |
| Basal diet+Leucine+Methionine | 42·3 | 161 | 16·1 | 2·63 |
| Basal diet+Lysine+Methionine | 49·7 | 175 | 17·5 | 2·83 |
| Basal diet+Leucine+Lysine+Methionine | 48·8 | 160 | 16·0 | 3·04 |
| Control diet (10% Casein protein) | 77·7 | 246 | 24·6 | 3·15 |

Each figure in table is a mean of six values. (P. Reddy, unpublished.)

the food has been prepared from raw material which has been eaten by man and animals from earliest times, the possibility of processing damage requires that tests on animals be carried out in order to assess the nutritive value of the product. Alternative methods of assessment have been proposed. An enzymic digestibility procedure has been advocated by Stahmann and his co-workers, and has been applied widely to leaf protein concentrates (Akeson & Stahmann, 1965). This procedure, involving digestion by pepsin followed by pancreatin, provides a measure of the digestibility of the protein in terms of the amount of amino acids released, and indicated that LPC samples prepared by Pirie and by the Chayen 'impulse-rendering' process had biological values lower than egg protein but higher than beef, casein, soybean, yeast, wheat, flour, gluten, zein and gelatin.

Broadly these estimates agreed with the results of animal tests excepting for lucerne products. The enzyme test indicated these to be equivalent to LPC from other sources, whereas it is known from animal tests that in fact lucerne LPC is inferior because of its associated growth depressing factor. Again, one might be misled by high analytical values for methionine content into thinking that LPC provided a nutritional sufficiency of this amino acid, but, as stated above, animal tests have shown that supplementation with methionine produces improved performance in rat diets. The danger of omitting a biological test of a practical type in which the actual performance in the animal is measured is very clear. Nutritive value is affected by many factors, digestibility and amino acid pattern being of very considerable importance. With some materials, however, and lucerne LPC is one, some other factor such as toxicity or amino acid unavailability may be the overriding one and in such cases the only true safeguard is the animal test.

The broad relationship between nutritive value and the content of organic sulphur and phenolic compounds, which has been remarked upon by Singh, is yet another example. The existence of this very interesting connection was revealed by comparative experiments using rat growth as the criterion of quality (Singh, unpublished).

**Mixtures of LPC with other protein sources**

Estimations of biological value and net protein utilization with rats have been carried out on diets in which the LPC was the sole source of protein. More practical evaluations have, however, been carried out with pigs (Barber *et al*, 1959; Duckworth *et al*, 1961) and with chicks (Duckworth & Woodham, 1961; Woodham, 1970), in which the LPC has been fed along with cereals and with groundnut (Table 11). The excellence of the material as a complement for these other protein sources has been amply demonstrated. In addition Miller (1965) has given practical human diets containing LPC to rats and showed that even without methionine supplementation the LPC mixtures

TABLE 11. Gross Protein Values (GPV) of groundnut–LPC mixtures

| Mixture | GPV |
|---|---|
| Groundnut (100) | 40 |
| Groundnut (66) – Oats LPC (33) | 60 |
| Groundnut (50) – Oats LPC (50) | 64 |
| Groundnut (33) – Oats LPC (66) | 65 |
| Oats LPC (100) | 66 |

S.E. of difference between means ±4·6.

were of good feeding value. Supplementation of rice with lucerne LPC gave excellent weight gains with rats, better even than those given by a skimmed milk powder supplement when each was included to provide 5% of protein. Liver protein levels in the supplemented diets were also increased.

Supplementation of wheat flour with LPC has also been extensively studied using rats, and from estimates of PER and liver nitrogen it was concluded that a mixture in which wheat and LPC provided equal amounts of protein was superior to mixtures containing other proportions, and that this superiority was attributable to the fact that the 50:50 mixture provided the best amino acid spectrum (Table 12). Mixtures of a standard rat diet with varying proportions of powdered milk or leaf protein were fed to rats over a 4-week period; the best mixture was one consisting of equal parts of basal, LPC and milk (Oke, unpublished). This mixture, in which 58% of the total protein was derived from LPC produced a weight gain of 55 g while the worst result was given by a 50:50 mixture of basal and LPC, the weight gain over the 4-week period being only 17 g (Table 13). Oke also fed rabbits on diets in which maize was progressively replaced by LPC. Combinations of maize with 10, 20 and 30% of leaf protein all gave better growth than maize alone, 10% replacement giving maximum weekly weight gain, but 30% having the greater apparent digestibility.

Recently Eggum (1970b) gave rats diets in which half of the protein was provided by cassava LPC (biological value=49), and half by dried cod (biological value=78). The biological value of the mixture was found to be 73 indicating that the LPC, though lower in BV than the fish, has an amino acid composition which complements the latter very satisfactorily.

TABLE 12. Supplementation of wheat with leaf protein—growth and PER studies on male rats. Wheat (W) and wheat+leaf protein (W: LP) mixtures were fed to weanling male rats over a period of 4 weeks. The average of eight rats in each group in Expt I and those of seven in Expt II are presented.

| Protein ratios | Dietary protein (%) | Diet intake (g) | Gain in weight (g) | PER | Liver N in dm (mg/g) | Liver N Total (mg) |
|---|---|---|---|---|---|---|
| EXPERIMENT I | | | | | | |
| W100 | 10·5 | 163 | 29·6 | 1·73 | 65·2 | 450 |
| W80 : 20LP | 10·7 | 194 | 45·0 | 2·16 | 69·7 | 560 |
| W70 : 30LP | 10·6 | 189 | 44·4 | 2·22 | 69·0 | 580 |
| W60 : 40LP | 10·8 | 213 | 53·8 | 2·34 | 72·3 | 650 |
| W50 : 50LP | 10·8 | 235 | 61·3 | 2·43 | 77·6 | 720 |
| EXPERIMENT II | | | | | | |
| W70 : 30LP | 11·8 | 193 | 56·3 | 2·41 | 70·1 | 560 |
| W50 : 50LP | 11·8 | 223 | 79·8 | 3·02 | 77·5 | 850 |
| W40 : 60LP | 11·8 | 214 | 74·8 | 2·97 | 77·5 | 775 |
| W30 : 70LP | 12·1 | 203 | 65·9 | 2·65 | 76·6 | 760 |
| W20 : 80LP | 11·7 | 203 | 61·8 | 2·59 | 75·0 | 730 |
| 100 LP | 10·9 | 162 | 35·3 | 1·98 | 72·0 | 540 |

Leaf protein used in the two experiments was from different batches of vegetation processed on different days. (N. Singh, unpublished.)

TABLE 13. Protein contents of diets and total weight gained by rats during experimental period

| Cage | Experimental diet: Standard diet | Leaf protein | Powdered milk | % protein derived from LP | Initial weight (g) | Weight gained (g) | Weight gained (% original weight) |
|---|---|---|---|---|---|---|---|
| 1 (Control) | 100 | — | — | — | 98 | 30 | 31 |
| 2 | 75 | 25 | — | 53 | 95 | 34 | 36 |
| 3 | 50 | 50 | — | 77 | 86 | 17 | 20 |
| 4 | 33 | 33 | 33 | 58 | 51 | 55 | 108 |
| 5 | 50 | — | 50 | — | 54 | 42 | 78 |
| 6 | — | 50 | 50 | 70 | 51 | 40 | 79 |
| 7 | — | — | 100 | — | 80 | 35 | 44 |

(O. L. Oke, unpublished.)

Recently, Coimbatore workers have published the results of growth studies with rats fed diets based mainly on rice and cholam, as eaten by low-income families in the locality, supplemented with either redgram dhal or LPC. The LPC supplemented diet gave better growth and protein intake than did either the unsupplemented or the redgram dhal supplemented diets, but the latter had the highest PER. The PER of the LPC supplemented diet was similar to that of the control. The authors concluded that amino acid composition data was needed in addition to feeding trials on man.

## Conclusion

Doubtless animal testing will continue as a control on new processing methods and to provide information on the acceptability of products made from hitherto unused plant species before such materials are recommended for incorporation into diets for man. Although the amount of this type of testing which has been carried out so far on leaf protein concentrates is not large, the results obtained are unequivocal and it can safely be claimed that the suitability of the material for the feeding of animals, and in particular for the non-ruminant, has been satisfactorily proved.

# 12

# Feeding Trials with Children

N. SINGH

In Jamaica, diets in which half the N was supplied by leaf protein were used in the rehabilitation of malnourished infants (Waterlow, 1962). At an ordinary level of N intake, retention was equal to that on milk alone fed at the same level; at a more generous level, there was better retention with milk. The essence of that paper is very briefly summarized in Table 1. These results, combined with those from feeding several species of animal in many different institutes, were the background to a long-term trial with children (Doraiswamy *et al*, 1969).

TABLE 1. Retention of N by infants recovering from malnutrition on diets containing milk (M) as sole source of N, and diets (LPM) with half the milk replaced by leaf protein

| Diet | M | LPM | M | LPM |
|---|---|---|---|---|
| No. of trials | 11 | 5 | 10 | 5 |
| N intake (mg/kg/day) | 776 | 765 | 504 | 496 |
| N retained (mg/kg/day) | 276 | 246 | 165 | 160 |

During three experimental periods of 12 days each 10–12-year-old children were given a low-protein diet (I), a diet containing uncooked LP (II), and one containing cooked LP (III). The faecal and urinary excretions during the last 4 days were collected for analysis. The low-protein diet with no LP (I) provided 12·6 g of protein/child/day and the uncooked (II) and cooked (III) LP-containing diets 45·0 and 48·8 g respectively. The LP contributed 72–74% of the dietary protein in the last two diets. Between the II and III periods, the children were kept for 9 days on a normal diet followed by 7 days on a low-protein diet. Leaf protein was distributed over the three meals, suitably incorporated into tapioca flour balls, vegetable curry and chutney. LP was cooked by boiling it in water and squeezing out the excess water before use.

TABLE 2. Metabolic studies with children: all values are per child per day as average of seven children

| Dietary period | N intake (g) | Excreted N | | | N Balance (g) | App. dig. (%) | N-retention: | | TD (%) | BV (%) |
|---|---|---|---|---|---|---|---|---|---|---|
| | | Faecal (g) | Urinary (g) | Total (g) | | | of ingested (%) | of absorbed (%) | | |
| Low-protein (I) | 2·05 | 0·85 | 1·15 | 2·00 | +0·05 | 58·5 | 2·5 | 4·2 | — | — |
| Uncooked LP (II) | 7·20 | 1·96 | 3·48 | 5·44 | +1·76 | 72·8 | 24·4 | 33·5 | 85 | 63·5 |
| Cooked LP (III) | 7·80 | 2·16 | 3·82 | 5·98 | +1·82 | 72·5 | 23·3 | 32·3 | 83 | 59 |

TABLE 3. Feeding trial with children. Growth and clinical responses in children on 6 month feeding with basal ragi diet and diets supplemented with lysine, leaf protein or sesame flour, and the intermediary N-metabolism in six paired children from each group

| | Dietary supplement | | | | |
|---|---|---|---|---|---|
| | None (1) | Lysine (2) | Leaf protein (3) | Sesame flour (4) | SEM |
| *Mean increases in*: (twenty children in each group) (57 df) | | | | | |
| Height (cm) | 2·20 | 4·25 | 4·84 | 3·51 | ±0·11 |
| Weight (kg) | 0·47 | 1·05 | 1·28 | 0·86 | ±0·04 |
| Haemoglobin (g/100 ml) | 0·29 | 0·64 | 0·87 | 0·73 | ±0·09 |
| RBC count ($10^{-6}$/mm$^3$) | 0·06 | 0·22 | 0·23 | 0·19 | ±0·02 |
| *General nutritional status*: (twenty children in each group) | | | | | |
| No. improved | 3 | 11 | 13 | 8 | |
| No. stationary | 11 | 9 | 7 | 11 | |
| No. deteriorated | 6 | 0 | 0 | 1 | |
| *Intermediary N-metabolism*: (six paired children from each group) (15 df) | | | | | |
| N-intake* | 6·10 | 6·32 | 7·75 | 7·83 | |
| Excreted N-faecal* | 2·74 | 2·52 | 2·63 | 2·79 | |
| Excreted N-urinary* | 2·61 | 2·30 | 3·23 | 3·43 | |
| Excreted N-total* | 5·35 | 4·82 | 5·86 | 6·22 | |
| N-balance* | 0·75 | 1·50 | 1·89 | 1·61 | ±0·15 |
| Apparent digestibility (%) | 55·1 | 60·2 | 66·0 | 64·4 | ±1·2 |
| N-retention as % of N ingested | 22·6 | 39·5 | 37·0 | 40·0 | |
| N-retention as % of N-intake | 12·6 | 23·7 | 24·4 | 20·6 | |

* All N values in g/child/day.

One child out of eight under study developed facial oedema, probably an allergy, on the 12th day of period II and the results, therefore, are presented as average of seven children (Table 2).

The results showed remarkable improvements in the quality of the low-protein diet on addition of leaf protein, as reflected in the digestibility and retention of N. Between the cooked and uncooked LP, there were no differences. In view of some allergic reactions in both Waterlow's studies and the present ones, it is reasonable to suggest that the quantity of leaf protein given should at present be smaller than that used here.

A 6-month long feeding trial with children on ragi-based diets was undertaken to compare the efficiency of a leaf protein compared to a lysine or a sesame flour supplement. Four groups of twenty children, 6–11 years old, were given per child per day: (1) basal ragi diet, (2) basal diet supplemented with 0·5 g of synthetic lysine, (3) basal diet supplemented with 15 g of leaf protein (providing 10 g of 100% protein), and (4) basal diet supplemented with 25 g of sesame flour (providing 10 g of 100% protein). The effects of such supplementation on the growth and nutritional status of children were studied, and midway during the feeding trial N-metabolism studies were done on six paired children from each group. The results are presented in Table 3.

Supplementation of ragi diets with any of the materials brought about improvements in all nutritional responses, i.e. height, weight, general nutritional status, apparent digestibility and N retention. The diet supplemented with leaf protein led to greatest growth response followed by those supplemented with lysine and sesame flour in that order. The apparent digestibility of the lysine-supplemented diet was less than that of the diets supplemented with leaf protein and sesame flour, while the differences in N retention in children between the three supplemented diets were not significant. However, the N-retentions, as % of ingested N on lysine and leaf protein supplements, were similar and better than on sesame supplement.

# 13

# The Use of the By-products from Leaf Protein Extraction

N. W. Pirie

If it is necessary to extract the lipids from leaf protein before it can be used extensively as a human food, the lipids will be a by-product. The carotene and xanthophyll should then be recovered, but other uses for this fraction need not be considered here because of the hope that extraction will not be necessary. The two by-products that must be considered are the protein-depleted fibre residue and the soluble uncoagulated components of the leaf.

In the late 1930s some firms in Britain connected with the processing of vegetables and the production of animal feedingstuffs examined the possibility of cheapening the production of dried cattle fodder by pressing some of the water out of herbage instead of evaporating it in the usual manner. They did not realize that it is difficult to press fluid out of an undamaged leaf. When cattle-raising became an important industry in Florida the late Dr Randolph was confronted with the problem of drying the local lush vegetation for use as a protein-rich supplement to the citrus waste that was the mainstay of the industry. A screw expeller (Casselman *et al*, 1965) was used to express the 'moisture'. In an expeller, leaves are pulped as well as pressed. Protein is therefore lost to an extent approximately proportional to the extent to which water is being removed. If the sole object of treatment is to remove water and thus make drying more economical, it would be better to heat the forage to 80° or 100°, and so coagulate the protein in it, and then press out a liquor that would carry with it little but the low-molecular weight components of the leaf (Pirie, 1966b). The thermal advantages of this are obvious; the material to be dried contains 25–35% DM whereas the raw forage contains only 10–15% which means that three times as much water has to be evaporated from it as from heated and pressed material. But economy and the avoidance of pollution make it essential that full use is made of the expressed liquor. The economics of the process are being examined in Britain and elsewhere.

The papers by Hollo & Koch and by Kohler & Bickoff stress the value of the fibrous residue as feed for ruminants. Its value is also recognized in New Zealand; a feeding experiment on sheep (p. 117) compared the fresh moist residue, and the same residue dried or ensiled. The fresh material was the most digestible, but the other two were readily accepted. In similar experiments with cattle (Oelshlegel *et al*, 1969b), silage made from the residue after protein extraction, from several plant species, was readily eaten and chemical analysis suggested that lucerne silage made from the residue was preferable to that made from the fresh crop. Experiments are in progress with dried and ensiled material at some other institutes.

The soluble but uncoagulable components of the leaf would have to be used to avoid local pollution. Hartman *et al* (1967) spray dry the whole extract. This procedure was condemned on theoretical grounds (Pirie, 1969b) and was shown experimentally to be harmful by Subba Rau *et al* (1969). Hollo & Koch (p. 65) coagulate and filter the extract and concentrate the filtrate in a multiple effect evaporator before adding it back to the protein. This is thermally efficient but does not seem to meet the objections. Kohler & Bickoff (p. 69) concentrate it and consider it valuable because of 'unidentified growth factors'. The merits of this material as an animal feedingstuff seem to be uncertain.

The uncoagulable material is a good culture medium for microorganisms. Its use is being studied in many parts of the world but most of this research is unpublished. The composition of this 'whey' depends on the species and maturity of the crop from which it came and, obviously, on the amount of water added either to improve extraction or in the course of washing the crop. At Rothamsted it is exceptional for it to have a DM outside the range 1·2–4%. Average values for the N and carbohydrate in the DM are 3 and 40%. The dominant monosaccharides are fructose and glucose. Polysaccharides account for about a quarter of the carbohydrate. In the grasses these are mainly fructosans but this is not so for all species, mustard, for example, contains little fructosan. It is important that there should be a more comprehensive set of analyses of the 'wheys' so that their efficient use can be planned.

Research on the 'whey' from pea-vines and other leafy material started in Uppsala about 1955. Using seven microorganisms (Jönsson, 1962), it was compared with various other substrates and proved particularly satisfactory for growing *Rhizobium meliloti*, and adequate for *Penicillium chrysogenum* and *Aspergillus niger*. More recently, in the Pakistan Council for Scientific and Industrial Research Laboratory (Lahore), Shah *et al* (unpublished) used extracts from three species. These had the compositions shown in Table 1. They were seeded with nine yeast strains, and centrifuged and dried after 48 hr shaking at 30° and pH 5. Table 2 records the yields. Growth was increased usefully when glucose and/or $(NH_4)_2SO_4$ was added to the *C. psoraliodes* extract but not when added to the others. Clearly, this group of yeasts is able to make quick use of only about half the organic matter in these fluids.

TABLE 1. Chemical composition of leaf filtrate

| | Dry weight (g/100 g) | Reducing sugar (g/100 ml) | Total nitrogen (g/100 ml) |
|---|---|---|---|
| *Trifolium alexandrinum* | 3·85 | 1·1 | 0·176 |
| *Phaseolus mungo* | 4·50 | 1·0 | 0·259 |
| *Cyamopsis psoralioides* | 1·11 | 0·7 | 0·107 |

TABLE 2. Growth of yeast on leaf filtrate in 48 hr (g dry matter per 100 ml)

| | *Trifolium alexandrinum* | *Phaseolus mungo* | *Cyamopsis psoralioides* |
|---|---|---|---|
| *Candida utilis* NCYC No. 193 | 0·72 | 0·58 | 0·50 |
| *Candida utilis* No. 359 | 0·62 | 0·51 | 0·40 |
| *Torula utilis* NRC No. 862 | 0·72 | 0·50 | 0·57 |
| *Saccharomyces cerevisiae* IMI'No. 399916 | 0·62 | 0·52 | 0·39 |
| *Rhodotorula* sp. ATCC No. 9058 | 0·59 | 0·29 | 0·25 |
| *Torulopsis mongolea* IFO No. 705 | 0·52 | 0·55 | 0·42 |
| *Candida guilliermondii* ATCC No. 9058 | 0·58 | 0·60 | 0·31 |
| *Debromyces subglobosus* NCYC No. 459 | 0·51 | 0·36 | 0·35 |
| *Candida robusta* IFO No. 735 | 0·33 | 0·34 | 0·29 |

# 14

# Carotene Content of Leaf Protein Preparations and their Use as Sources of Vitamin A

A. Pirie

Many people in many parts of the world rely on vegetables and fruit to supply vitamin A through the pro-vitamin, β-carotene. But they, particularly the young children, do not get enough. From the point of view of health and well-being and to prevent blindness it is important to increase the intake of vitamin A. Without it children fail to grow, they become listless and their vision and eyes show characteristic changes. First they become night-blind, for there is no vitamin A to form rhodopsin, one of the light-sensitive pigments of the retina. Next their conjunctivas and corneas become dry and lustreless (xerophthalmia) and then the central cornea becomes opaque and may ulcerate and even perforate (keratomalacia) so that the eye is lost and the child blind (Oomen, 1961).

The vitamin itself is present only in animal foods such as milk, butter, liver, meat and eggs, but carotenes, in particular β-carotene, which is widely present in green leaves and some roots and fruits, can be converted to vitamin A in the gut and after absorption is then stored in the liver. Deficiency may arise because the intake is too small or because the child is not absorbing the vitamin from the gut, or because his needs are abnormally high owing to some illness. Absorption from the gut is impaired on a low protein or low fat diet. Kwashiorkor is often associated with vitamin A deficiency. Oomen (1961) says 'the association of xerophthalmia with protein malnutrition is so frequent and so close that we cannot help wondering which is the primary lesion'.

W.H.O. (1967) has assessed the biological activity of β-carotene in the human diet as one-sixth of that of an equal weight of vitamin A. On this basis they suggest that a child below 7 years old needs 1·8 mg β-carotene per day while an adult needs 4·5 mg. This may overstate the need; Indonesian children showed no signs of vitamin A deficiency and grew well with 1·1 mg β-carotene in their food (Blankhart, 1967). Seventy per cent of the carotene in 40 g of amaranth per day was absorbed from the gut by underweight

Indian children (Lala & Reddy, 1970) and by young healthy Indian men (Rao & Rao, 1970). This suggests that the biological value of β-carotene is more like one-quarter that of vitamin A rather than only one-sixth, as suggested by W.H.O. There is evidence that the conversion of β-carotene to vitamin A in the serum is more efficient if given as a vegetable than as β-carotene in oil. Hume & Krebs (1949) consider this may be because vegetables contain vitamin E and other naturally protective substances in close association with the carotene. The fineness of emulsion of carotene in natural foods is probably also a factor increasing its usage.

## The quantity and stability of β-carotene in leaf protein

It is easy to measure the amount of β-carotene in leaves and leaf protein because the hydroxy compounds (e.g. xanthophyll or lutein) are retained by adsorbents such as alumina, and there is too little of the other carotenoids (e.g. α-carotene) to interfere with measurements of the precision aimed at. Carotene is remarkably stable in the extracts used to make leaf protein on a large scale and during its subsequent processing (Arkcoll & Holden, 1971). About a third (9–42% in different leaves) of the carotene is extracted and precipitates with the protein without loss when the extract is heated to 80°. Fresh freeze-dried samples of leaf protein have very similar levels of β-carotene; from seven out of ten species it fell between 1·4 and 1·7 mg/g. Accepting the W.H.O. estimate of 1·8 mg/day needed by a young child, it is clear that 2 g leaf protein per day would more than suffice.

Pulping leaves caused no loss of carotene and little disappeared on storage of the pulp overnight at pH 7 though there was considerable loss at pH 4·5. Carotene was stable in leaf extract for at least 2 hr in the shade and there was little loss when the wet cake of protein was dried frozen but air drying at 40° caused a loss of about 30%.

Autooxidation causes considerable losses of β-carotene during storage. These can be minimized by drying at a low temperature and storing in the cold in the absence of light and oxygen. Oxidation of carotene was especially rapid in light. After 1 month's storage of freeze-dried leaf protein at 20° in polythene containers permeable to light and oxygen, 54% of β-carotene was lost, but only 25% was lost after 1 month in aluminium foil containers (not vacuum packed) which exclude light and oxygen. There was negligible loss on storage in the dark at −20° in either container.

Although these losses appear large they do not seriously affect the value of leaf protein as a source of vitamin A. The protein tested contained, before storage, 1·57 mg β-carotene per g so that 1 g can almost supply a child's need of vitamin A for 1 day. If half is lost, the intake of leaf protein must be doubled to 2 g. It is likely that at least this intake will be aimed at to counter protein deficiency. The presence of β-carotene in leaf protein, even after storage, can be regarded, at the very least, as a most important nutritional bonus.

## Use of leaf protein as source of vitamin A in the diet

Singh (unpublished) reported two experiments both of which show that leaf protein can be used as sole source of vitamin A in the diet of rats. Weanling rats were fed fresh wet lucerne leaf protein as sole source of vitamin A for 6 weeks. They grew as well as, and had slightly higher serum and liver vitamin A levels than, similar rats fed an equivalent diet using synthetic β-carotene as vitamin A source (Table 1).

In the second experiment (Table 2) young rats were kept on a vitamin A free diet until their weight plateaued or fell. For 18 days they were then given synthetic β-carotene or lucerne leaf protein as sole source of vitamin A. In both groups the β-carotene level was 680 μg/100 g diet and total protein 18%. The young rats fed lucerne leaf protein did better in every respect than those

TABLE 1. Studies on lucerne leaf protein as a source of β-carotene for young rats

| Diet | Initial body weight (g) | Total diet intake (g) | β-carotene intake (μg/day) | Total gain in weight (g) | Liver: Fresh weight (g) | Liver: Total vit. A (μg) | Serum vit. A (μg/100 ml) |
|---|---|---|---|---|---|---|---|
| A. Control | 37·9 | 336 | 96 | 138 | 7·3 | 400 | 59 |
| B. Test | 37·9 | 358 | 103 | 141 | 6·9 | 451 | 68 |

Diets containing 1220 μg β-carotene/100 g and 18% protein were fed to weanling (21-day-old) rats *ad libitum* over a period of 6 weeks. Control diet had only synthetic β-carotene (E-Merck) and total protein from casein (vitamin A free). Test diet had all β-carotene from leaf protein (5 g wet cake) which also provided some protein; casein made up the protein to 18%. All values are the mean of eleven rats. (N. Singh, unpublished.)

TABLE 2. Studies on lucerne leaf protein as a source of β-carotene for rehabilitation of young rats with early symptoms of vitamin A deficiency

| Diet | Initial weight (g) | | Diet intake (g) | | β-carotene intake (μg) | | Gain in weight (g) | | Liver fresh weight (g) | | Total liver vit. A (μg) | | Liver vit. A (μg/100 μg β-carotene intake) | | Serum vit. A (μg/100 ml) | |
|---|---|---|---|---|---|---|---|---|---|---|---|---|---|---|---|---|
| | M | F | M | F | M | F | M | F | M | F | M | F | M | F | M | F |
| Control | 110 | 107 | 146 | 138 | 996 | 935 | 39 | 23 | 6·1 | 4·9 | 52 | 70 | 5 | 7 | 50 | 47 |
| Test | 118 | 111 | 196 | 158 | 1330 | 1076 | 57 | 30 | 7·3 | 5·2 | 91 | 97 | 7 | 9 | 68 | 49 |

Young rats, showing early symptoms of vitamin A deficiency (loss in weight or its maintenance over a few days at around 47 days after weaning), were fed for a period of 18 days isogenous diets supplemented with either leaf protein (Test) or synthetic β-carotene (Control) to provide 680 μg of β-carotene for every 100 g of the diet. The means for five male and six female rats are presented (Singh, unpublished).

M=male; F=female.

on synthetic β-carotene. In the first place they obviously found the diet more palatable as they ate more than the controls. They gained considerably more weight and had higher serum and total liver vitamin A. Some of these gains may have been due to higher food intake but liver storage, when calculated as μg of vitamin A in the liver per 100 μg of β-carotene ingested, was higher in the leaf protein fed rats. In both these experiments the level of β-carotene in the diet was very high but as the main use of leaf protein in human food will be to prevent protein deficiency it is likely that the accompanying level of β-carotene will also be high in these circumstances.

# Section IV

# Presentation and Acceptability

# 15

# Acceptability of Food Preparations Containing Leaf Protein Concentrates

G. Kamalanathan and R. P. Devadas

Diet surveys carried out recently by Devadas & Easwaran (1967) and the Indian Council of Medical Research (1969) indicate the persistent preponderance of cereals in the Indian dietaries; this makes them ill-balanced, and often limits the protein intake. Various measures are being suggested to make up the protein gap, such as increased production of protein-rich foods and formulation of protein-rich mixtures from indigenous sources (Swaminathan, 1967). In order to meet the expanding needs for protein of the ever-increasing world population, all possible sources for protein need to be tapped (Pirie, 1969c).

Nutritionists in different parts of the world are investigating the possibilities of supplementing diets with novel sources of protein. One such source is Leaf Protein Concentrate. In order to gain acceptance for novel foods by those who need them most, special efforts need to be taken towards their popularization. However nutritious or abundant a novelty is, it may not be adopted unless its palatability has been established (Devadas, 1967). Therefore consumer acceptance through persuasion, education and example is of paramount importance when introducing novel foods.

Very few studies have been done on the incorporation of LPC into food preparations for human use. Fewer still are the data available on the acceptability of foods preparations with LPC. At Sri Avinashilingam Home Science College, studies on acceptability have been conducted since 1967, using freeze-dried LPC from Rothamsted Experimental Station, U.K., and air-dried or acetic acid preserved samples from the Central Food Technological Research Institute, Mysore.

**Wet and dry LPC**

Preliminary trials on treatment of wet and dry LPC to remove the strong flavour, indicated that repeated washing, or passing steam through the LPC for 5 min, helped to remove some of the 'grassy' flavour. Pressure cooking and roasting were also tried but they were not effective. Since dry LPC has the advantage of being compact and easily transportable requiring less space for storage, and longer shelf life when compared to wet LPC, all our subsequent studies have been conducted with dry LPC (Kamalanathan *et al*, 1969).

**Selection of the food preparations for the incorporation of the LPC**

As a basis for selection of recipes, which are popular in local dietaries, a survey was conducted among 100 households in Coimbatore city. It was found that the preparations and adjuncts occurring in the menu frequently were: 'Ragi addai', 'Dhal balls', 'Sweet potato curry', 'Potato bath', 'Leaves chutney', 'Chutney powder', 'Sambar bath', 'Brinjal porial', 'Greens kolumbu' and 'Pumpkin porial'.

Among these, six items were chosen to present the main dishes or adjuncts, and various cooking methods as indicated in Table 1.

TABLE 1. Details of the six selected food preparations

| Food preparation | Method of cooking employed | Consumed as: |
|---|---|---|
| 1. Dhal balls | Steaming | Side dish |
| 2. Chutney powder | Roasting | Adjunct |
| 3. Sweet potato curry | Boiling | Side dish |
| 4. Ragi addai | Shallow fat frying | Main dish for breakfast |
| 5. Potato bath | Simmering | Main dish for breakfast or lunch |
| 6. (Leaves) chutney | Consumed raw (ingredients are ground together) | Adjunct |

**Standardization of the recipes**

Amerine *et al* (1965) suggest that a standardized recipe should be such that it produces identical results whenever tried under the conditions specified. Accordingly, all the variables such as the ingredients, cooking temperature, duration of cooking, the quantity of water, salt, and condiments used and blending were controlled.

For arriving at the quantity of the LPC to be used, from the nutritional point of view, it was considered advisable that the LPC furnishes 6–7 g of protein; for this 10–13 g of the normal product is incorporated. Hence recipes were developed in which 10 g of LPC was incorporated at each serving, and at two other levels, one lower, namely 5 g, and one higher, 15 g, per serving.

For purposes of comparison, a 'Standard' recipe without LPC was used. Appendix I gives the recipes.

Five judges were selected through methods accepted for selecting the taste panels. A score card on a five-point scale ranging 1 to 5 was used for facilitating the tasting by the judges. It itemized colour, taste, texture and flavour. The recipes were prepared on five different occasions and served under identical conditions to the five judges. The mean score was thus the mean of the scores given by five judges on five occasions (that is, the mean of twenty-five replicates).

**The scores awarded for the recipes**

The mean scores awarded to the different food preparations with and without LPC are presented in Table 2.

TABLE 2. Mean total scores of the LPC incorporated products and the standards (maximum possible score = 20)

| S. No. | Products | Standard without LPC | LPC per serving | | |
|---|---|---|---|---|---|
| | | | 5 g | 10 g | 15 g |
| 1. | Dhal balls | 18·20 | 18·52 | 11·44 | 9·56 |
| 2. | Chutney powder | 17·48 | 13·88 | 12·92 | 12·20 |
| 3. | Sweet potato curry | 16·20 | 10·92 | 10·08 | 8·72 |
| 4. | Ragi addai | 16·24 | 13·52 | 11·24 | 11·00 |
| 5. | Potato bath | 18·20 | 12·76 | 11·28 | 9·44 |
| 6. | Leaves chutney | 18·48 | 16·04 | 11·40 | 12·16 |

At 5 g level, the scores of the dhal balls, leaves chutney, chutney powder and ragi addai are close to the standard. At the higher level of 15 g all the samples were reported to have leafy flavour, saw-dust texture and bitter taste. The 10 g was intermediate between the two and still acceptable having secured more than 60% of scores awarded to the standards.

An analysis of the scores with regard to the individual attributes of colour, texture, flavour and taste indicated that:

(a) The colour is more acceptable in preparations which are normally green or dark in colour.

(b) Texture is not so much a problem as LPC combines uniformly with the other ingredients.

(c) Grassy flavour is a problem in preparations which do not have strong spices or bland bases such as banana.

(d) The marked bitter or sour taste is a problem.

This study further showed that the LPC was more acceptable when introduced at the completion of cooking, and when the strong flavours were masked by spices or fully ripe bananas.

An analysis of the scores awarded by the judges on different occasions for the same preparation over a period of 5 months indicates a trend that as each judge got used to the LPC products, the reaction was more favourable in the succeeding scorings. For instance, with regard to chutney powder, while the mean score the first time it was presented was 10/20, at the fifth scoring it went up to 14/20.

**Leaf protein in flour products**

In another series of experiments, LPC incorporated at 5% level in Sujji (fine broken wheat) and Maida (refined wheat flour) sent from CFTRI, were tested in simple recipes such as uppuma for sujji, and diamond cuts for maida. A group of thirty graduate women tasted the products and gave the scores presented in Table 3. Both uppuma and diamond cuts were accepted well.

TABLE 3. Mean scores for LPC uppuma and diamond cuts

| Quality | Max. score | Uppuma | | Diamond cuts | |
|---|---|---|---|---|---|
| | | Standard | With LPC | Standard | With LPC |
| Colour | 50 | 49 | 41 | 50 | 37 |
| Texture | 50 | 46 | 50 | 48 | 50 |
| Flavour | 50 | 41 | 36 | 47 | 46 |
| Taste | 50 | 41 | 29 | 42 | 47 |
| Total | 200 | 177 | 156 | 187 | 180 |

**Feeding experiments with albino rats**

Devadas *et al* (1970) carried out a study on albino rats to evaluate the effects of LPC supplementation on the nutritive quality of the diets of a selected group of rural families, whose monthly income ranged from Rs. 100 to 200. Three diets were used in this study:

Diet A—Basal diet in which the protein content was 9·0% (Appendix 2).

Diet B—Basal diet plus 5% protein from redgram dhal.

Diet C—Basal diet plus 5% protein from LPC (extracted from lucerne).

Their findings revealed that:

(a) The supplementation of the basal diet with redgram dhal and LPC significantly increases the growth rates, in spite of the fact that the rats receiving the basal diet alone, ate more than the other groups.

(b) The rats on diet B ate more protein than those on diet A and less than those on diet C.

(c) The PER values of groups A and C were equal and significantly less than that of group B.

(d) The hepatic N contents of groups B and C were equal and greater than that of group A.

**Feeding trial on children with LPC Laddu**

Our experience in village work (Kamalanathan *et al*, 1970) shows that in order to enhance the intake of protein, the novel food needs to be prepared with the following points in view:

(a) Supply maximum protein in a concentrated form.
(b) The preparation should be a ready-to-eat product.
(c) The dish must be convenient for distribution.
(d) The dish must be palatable and easy to consume.

A preparation which approaches these conditions was formulated in our laboratories and given the name 'Laddu' because it resembles this popular sweet preparation, used for festivals and feasts and hence possessing 'prestige' value. The composition of the Laddu is given in Table 4 and the recipe in

TABLE 4. Composition, cost and protein value of laddu

| Ingredients | Weight (g) | Dry wt (g) | Protein (g) | N content (g) | Cost Paise |
|---|---|---|---|---|---|
| Maize roasted | 30 | 25·5 | 3·3 | 0·533 | 2·4 |
| Greengram dhal | 20 | 18·0 | 4·9 | 0·784 | 3·9 |
| Groundnut | 10 | 9·6 | 3·2 | 0·504 | 3·5 |
| Jaggery | 30 | 28·8 | 0·1 | 0·019 | 2·0 |
| LPC | 10 | 10·0 | 6·0 | 0·960 | 1·0 |
| Total | 100 | 91·9 | 17·5 | 2·800 | 12·8 |

N, as % of dry matter (calculated)=3·45.
% of N contributed by leaf protein=34.

Appendix 3. As the ingredients are precooked and can be easily mixed with water and made into balls, the mothers can give the food to their children without hesitation.

A hundred grams of Laddu on dry weight basis supply 17·5 g protein. This amount of protein is 50% of the recommended allowance for the age group 7–9 years (ICMR, 1968). Laddu incorporating 10% LPC is well accepted and 100 g of Laddu can be eaten easily by a child. The greatest advantage of this recipe is its low cost.

A longitudinal study has just been started to determine the effect of feeding Laddu incorporating LPC. Five mothers from families of low socio-economic status having children in the age range of 6–10 years, were approached and informed of the availability of a special food which is 'green' and good for health. A dietary survey by the oral questionnaire cum food volume measurement method was conducted to elicit the pattern and quantity of food intake. Thereafter, ten boys, 6–8 years of age, were selected, who were receiving home diets which were short of the required quantity of protein by 18–20%, that is by 6 g. Their height, weight and haemoglobin level were recorded and a clinical assessment was made.

TABLE 5. Comparative cost of different sources of protein

| Foodstuffs | Cost per 100 g (Paise) | Protein (g/100) | Cost to fill the gap of 6 g protein (Paise) |
|---|---|---|---|
| *Vegetable sources* | | | |
| Sesame | 35 | 31·5 | 6·6 |
| Groundnut | 35 | 31·5 | 6·6 |
| Wheat flour | 12 | 12·1 | 6·0 |
| Ragi | 6 | 7·3 | 4·8 |
| Redgram dhal | 18 | 22·3 | 4·8 |
| Maize | 8 | 11·1 | 4·2 |
| Bengalgram dhal | 14 | 17·1 | 4·2 |
| Greengram dhal | 19 | 24·0 | 4·2 |
| LPC (estimated by N. Singh) | 10 | 60·0 | 1·0 |
| *Animal foods* | | | |
| Milk | 11 | 3·0 | 21·6 |
| Egg | 60 | 18·3 | 20·0 |
| Meat | 60 | 18·5 | 19·8 |

LPC is the least expensive source of protein.

The investigators invited the mothers and the children to taste the Laddu, which they themselves consumed in the presence of the mothers. After ensuring acceptability, the experiment started.

Table 5 is an analysis of the cost of LPC Laddu in comparison with that of other foods furnishing the same quantity of protein.

A continued and plentiful supply of LPC is needed for further studies of its long-term acceptability and contribution to the protein calorie and carotene content of diets. Ultimately it should be on the market.

## APPENDIX 1

In all these recipes, pressed leaf protein containing about 70% of water is used.

1. *Dhal balls*

| *Ingredients* | *Wt (g)* |
|---|---|
| Redgram dhal | 28·5 |
| Coriander leaves | 1·0 |
| Curry leaves | 1·0 |
| Green chillies | 1·0 |
| Onion | 8·0 |
| Salt | 0·5 |

*Method:* Soak the dhal for 30 min. Grind into a coarse paste. Add chopped coriander leaves, curry leaves, onion and salt. Mix well and form into lime-sized balls. Steam them for 15 min. Yield: One serving. With incorporation of LPC, replace part of the dahl with 6, 10 or 14 g.

2. *Chutney powder*

| *Ingredients* | *Wt (g)* |
|---|---|
| Bengalgram dhal | 25 |
| Blackgram dhal | 12 |
| Sesame | 3 |
| Tamarind | 5 |
| Curry leaves | 1 |
| Coriander leaves | 1 |
| Red chillies | 5 |
| Jaggery | 2 |
| Salt | 3 |
| Copra (desiccated coconut) | 7 |

*Method:* Dry roast all the ingredients except jaggery, salt and copra separately, and mix. Dry roast copra and add to the mixture. Add powdered jaggery and salt. Pound into a medium coarse powder. Yield: Two servings. Approximate weight per serving: 20 g. With incorporation of LPC replace part of the Bengalgram dhal with 13·5, 20 or 23 g.

3. *Sweet potato 'curry'*

| *Ingredients* | *Wt (g)* |
|---|---|
| Sweet potato | 75 |
| Salt | 3 |
| Coriander<br>Clove<br>Cinnamon powder | 2 |
| Onion | 9·5 |
| Oil | 10 |
| Mustard | 0·5 |

*Method:* Steam the sweet potatoes until cooked. Peel and dice them. Make a seasoning of mustard in a pan. Add the chopped onion. When the onion is cooked, add the spices and the diced sweet potato and mix well. Sprinkle one or two teaspoons of hot water if needed. Yield: One serving of approximately 75 g. With incorporation of LPC substitute 5, 10 or 15 g for part of the sweet potato.

4. *Ragi 'addai'*

| *Ingredients* | *Wt (g)* |
|---|---|
| Ragi flour | 100 |
| Ripe banana | 50 |
| Jaggery | 30 |
| Oil | 10 |

*Method:* Steam ragi flour for 15 min. Mash the banana and dissolve powdered jaggery in it. Add the steamed ragi flour and mix thoroughly. Add a little water, if necessary, and knead into a soft dough. Grease a hot iron shallow frying pan and flatten half of the dough into a thin 'addai' on it. Add oil around the edges and cook, turning it over when one side is cooked. Yield: Two servings of approximately 100 g each. With incorporation of LPC replace part of the ragi flour with 10, 20 or 30 g.

5. *Potato 'bhath'*

| *Ingredients* | *Wt (g)* |
|---|---|
| Ravai | 50 |
| Potato | 50 |
| Green chillie | 1 |
| Ginger | 0·25 |
| Turmeric powder | 0·25 |
| Masala powder* | 2 |
| Oil | 10 |
| Curry leaves | 0·5 |
| Salt | 5 |

| * *Masala powder* | |
|---|---|
| Cumin seeds | 10 |
| Coriander seeds | 10 |
| Cardomom | 2 |
| Cinnamon | 2 |
| Clove | 2 |

To be lightly dry roasted and powdered

*Method:* Boil potato, peel and dice. Heat oil in a pan and add curry leaves, ginger, turmeric powder and masala powder. After the curry leaves become crisp, add the ravai and roast until the ravai smells fragrant. Boil 300 ml (about 10 oz) of water in a vessel, and add the potato pieces, salt, and ravai, and mix well. Lower the heat and cook on slow heat till done, with constant stirring. With incorporation of LPC, replace part of the ravi with 15, 20 or 30 g.

6. *Leaves 'chutney'* (Veppilaikatti)

| *Ingredients* | *Wt (g)* |
|---|---|
| Bitter lemon leaves (tender) | 30 |
| Green chillies | 15 |
| Tamarind | 25 |
| Coriander seeds and cumin seeds | 4 |
| Salt | 5 |
| Oil | 10 |
| Mustard | 1 |

*Method*: Grind all the ingredients except oil and mustard together into a fine paste and season with mustard. Yield: Three servings of approximately 30 g. With incorporation of LPC, add 15, 30 or 45 g.

# APPENDIX 2

THE COMPOSITION OF THE BASAL DIET A

| *Foodstuff* | *Quantity* (*g*) |
| --- | --- |
| Rice (parboiled, milled) | 39 |
| Cholam | 33 |
| Redgram dhal | 7 |
| Horse-gram dhal | 6 |
| Amaranth | 2 |
| Brinjal | 3 |
| Potato | 6 |
| Mutton | 2 |
| Milk | 1 |
| Groundnut oil | 1 |
| Total | 100 |

# APPENDIX 3

*Laddu*

| *Ingredients* | *Wt* (*g*) |
| --- | --- |
| Maize | 30 |
| Greengram dhal | 20 |
| Jaggery | 30 |
| Groundnut | 10 |
| LPC | 10 |

*Method:* Prepare flour out of maize, green gram dhal and roasted groundnut. Mix with jaggery, and LPC. Add 15 ml of hot water and prepare balls.

# 16

# The Presentation of Leaf Protein on the Table

N. W. PIRIE

It is not easy to lay down rules for acceptability trials. Obviously, a main food about which traditions have built up should not be interfered with. During both the world wars there were protests in Britain when the percentage extraction of the flour used for bread was increased; bread would be an unsuitable vehicle for anything that will affect its colour. If a food is being judged as a good or bad example of a familiar type, it is extremely unlikely that any addition, at the 3–10% level, will be found other than detrimental. The question at issue is not 'Is this familiar food as good as usual?' it is rather 'Would this food be acceptable to people who understood the need for a protein supplement?' There is little chance of getting a novelty such as leaf protein into use on a significant scale by stealth. As with other changes in food habits, the change will not be made till people see a real or imaginary advantage in the change.

The larger the amount of protein used in an experiment to assess its nutritional merits, the more quickly and positively the assessment will be made. Except in infant feeding, it is both unusual and undesirable that people should get most of their protein from a single source. Furthermore, both Waterlow (1962) and Singh (p. 133) encountered what were thought to be allergic reactions to leaf protein. It may be that these reactions were a coincidence and had nothing to do with the supplement. They may have been reactions to a contaminant such as mould on the leaf surface that can be avoided once recognized. They may have been the result of photosensitivity brought on by uncommon intestinal bacteria acting on chlorophyll and its breakdown products and making phylloerythrin which is known to sensitize animals (Aronoff, 1953), or to the accumulation of phytanic acid (Refsum's disease). There are other possibilities. Most people are sensitive to some type of food. Until the point is cleared up, the amount of leaf protein eaten per day should probably not exceed 10 g in adults and correspondingly less in children. That is more protein than is eaten as meat in most countries, or as fish in any country except Portugal. The amount used should not, on the other hand, be unrealistically small. Nothing is gained by a token addition.

Some of the principles underlying presentation were outlined by Byers *et al* (1965). It may be worth reprinting part of that paper and some of the recipes given in it:

The attitude of the people for whom we cook is, naturally, somewhat different from that of experienced cooks. Those who regularly see leaf protein accept its appearance, and no particular finesse is needed in presentation; this would probably soon be the attitude of mind in any community in which it was being made and used, and in such a community a more varied group of dishes would be produced than we have any need for. Our problem is, in the long run, unreal; it is to make something sufficiently attractive to appeal to visitors—especially those visitors who are responsible for the agriculture and food policy of an international agency, grant-giving body, or developing country. Until we have interested these people nothing can be done to meet the simpler requirements of the needy. This however is not the sole criterion of a satisfactory form of presentation.

Leaf protein is intended as a protein supplement. Expressed in terms of dry matter, an adequate diet for those whose protein needs deserve most attention —those who are growing, pregnant, or lactating—should contain 15–20% protein. It is not therefore worth while working on foods that are palatable only when they contain less protein than this, and we have concentrated attention on dishes in which leaf protein makes up about half of the total. Because it is desirable that the protein in a diet should come from many sources, no attempt has been made to introduce enough leaf protein to supply the full requirement for a day; a reasonable helping of one of the dishes described here contains 6–10 g; a tenth of the total desirable daily intake.

The simplest method of presentation is as good as any, but it will only be used by people who have become accustomed to the idea of leaf protein. In this method crumbled moist protein is sprinkled, at the table, on to a risotto or some similar fairly highly flavoured but protein-deficient dish. The unblended particles of protein are by no means unattractive so long as one has added them oneself; it is more difficult to get acceptance if the mixture is made in the kitchen. In some dishes the protein is made into pieces encased in thin batter or pastry that are eaten without being bitten and examined, or that are bitten only once. For obscure reasons people seem to be much less concerned about the internal appearance of small pieces of food than they are about large ones. This can readily be confirmed at cocktail parties or in an Indian restaurant. We assume, however, that the appearance that people accept as normal in a food is purely a matter of convention and that most will accept the somewhat novel appearance of our products after they become familiar with them. Similarly, people habituated to leaf protein accept its flavour so that a larger proportion can be added to a food. Devadas has already commented on this (p. 148) and it was also the experience of Reddy (unpublished) who got panels to judge six traditional Indian dishes to which 3 and 6% of leaf protein had been added. Not unexpectedly, the dishes were judged to have suffered, but they were still acceptable—more so on second or third trials.

Texture is as important as flavour or colour. Freshly made slabs of leaf protein disperse in water to give a smooth paste but slabs stored in deep-freeze gradually become gritty and have to be passed through a mill, or ground in a mortar, before use. Material that has been skilfully freeze-dried disperses to a smooth paste with water. In general, savoury, or fish, blend better with leaf protein than sweet flavours or coffee, orange or lemon. An exception to this is banana.

A representative sample of each dish was freeze-dried, ground and analysed for dry matter and total N by Kjeldahl. The weight of N contributed by each component in the dish was either determined by analysis, or taken from food tables, or both. These figures enable an estimate to be made both of the final N content to be expected for the food and also of the proportion of that N present as leaf protein. Conventionally the N content of a foodstuff, multiplied by some factor such as 6·25, is taken as a measure of its protein content; this practice is not significantly misleading with leaf protein because the amount of NPN in our preparations (mainly lipid N) is only about 1% of the total. With the other ingredients of these dishes, much of the N is often present in forms other than protein. We have made no correction for this; consequently, leaf protein is in fact contributing a larger percentage of the total protein than we claim.

Our recipes are of three basic types: (1) a dry stable food that can be demonstrated to unexpected visitors and at scientific meetings; (2) mixtures enclosed in a case of batter or pastry so that the food, after cooking, has a perfectly conventional appearance; (3) soups and stews prepared in a conventional way but with leaf protein added just before serving. Representative recipes of each type are given together with the appropriate analyses, calculations, and comment on the flavour and limitations imposed by that form of presentation.

## Recipes and results

In all the recipes, the weight of freeze-dried protein is given because it is the material that those who are not making protein themselves are most likely to use; fresh protein is two-thirds water so three times as much would be taken, and less water, to get a similar result.

### Curry cubes

*Method.* The flour was mixed with a little water to a cream, and boiling water, in which a bag of mixed herbs had been boiling for ½ hr, was added slowly to the cream. This was cooked to make a smooth viscous sauce. Curry powder was added to fried chopped onion and, after further frying, was added to the sauce. Finally sodium glutamate and leaf protein were added. After thorough mixing the paste was frozen in trays half an inch deep and freeze-dried. The cooked flour holds the mass together so that it can be cut into cubes; they keep at room temperature for several weeks and indefinitely in the refrigerator. This mixture is designed for people who like a fairly strong curry.

| Contents | Wt taken (g) | Dry wt (g) | N content (g) |
|---|---|---|---|
| Flour | 28·0 | 24·40 | 0·479 |
| Onion | 24·0 | 1·73 | 0·036 |
| Curry powder | 10·0 | 10·00 | 0·152 |
| Margarine | 20·0 | 17·25 | 0·006 |
| Sodium monoglutamate | 4·5 | 4·50 | 0·336 |
| Barley leaf protein (freeze-dried) | 20·0 | 19·10 | 2·040 |
| Water | 276·0 | — | — |
| | 382·5 | 76·98 | 3·049 |

N as % of dry matter: calculated=3.96; by analysis=4·23. % of N due to leaf protein= 67·0. There is no need to add so much margarine. When less is added the final protein concentration is obviously greater.

### Banana mixture in batter

*Method.* An egg, flour and water batter was made by beating all the ingredients together; this gives a batter of 'coating consistency'. The filling, made by mashing the banana and then adding the leaf protein, was rolled into small balls, coated with batter, dropped into deep fat and fried for a few minutes until golden brown. Very ripe fruit is preferable as it has a stronger flavour than when partially ripe. This ratio of banana to protein successfully masks the flavour of most batches of protein. More protein can be used when people are accustomed to the flavour or when the preparation is exceptionally bland. A small amount of water for mixing is sometimes necessary if a high proportion of freeze-dried protein is used.

| Contents | Wt taken (g) | Dry wt (g) | N content (g) |
|---|---|---|---|
| (a) Flour | 112·0 | 97·90 | 1·915 |
| Egg | 59·5 | 15·80 | 1·130 |
| Water | 120·0 | — | — |
| | 291·5 | 113·7 | 3·045 |
| (b) Batter (see (a)) | 29·0 | 11·30 | 0·303 |
| Banana | 24·0 | 7·04 | 0·043 |
| Barley leaf protein (freeze-dried) | 4·0 | 3·82 | 0·408 |
| | 57·0 | 22·16 | 0·754 |

N as % of dry matter: calculated = 3·40; by analysis = 3·27. % of N due to leaf protein = 54·2.

**Banana fritters**

*Method.* Freeze-dried barley leaf protein was blended with water to approximate to the wet product (in proportion of 1:3). Equal weights of mashed banana and wet protein were mixed together and the filling put into thinly rolled pastry cases. These envelopes were dropped into deep hot fat and fried golden brown.

| Contents | Wt taken (g) | Dry wt (g) | N content (g) |
|---|---|---|---|
| (a) Banana | 100·0 | 29·3 | 0·18 |
| Barley leaf protein (freeze-dried) | 30·0 | 28·6 | 3·06 |
| Water | 70·0 | — | — |
| | 200·0 | 57·9 | 3·24 |
| (b) Banana filling (see (a)) | 71·0 | 20·5 | 1·15 |
| Pastry | 50·0 | 43·0 | 0·62 |
| | 121·0 | 63·5 | 1·77 |

N as % of dry matter: calculated = 2·79; by analysis = 2·28. % of N due to leaf protein = 61·5.

**Apple, onion and leaf protein pie**

*Method.* A thick puree was prepared by cooking the apples without water. The finely chopped onions and the sugar were added to this, followed by the leaf protein. The well-beaten mixture was put into a pastry-lined tin and covered with a pastry lid; the pie was cooked at 232° for 25 min. It can be served hot or cold.

| Contents | Wt taken (g) | Dry wt (g) | N content (g) |
|---|---|---|---|
| Cooking apple | 387·0 | 55·8 | 0·194 |
| Onion | 177·0 | 12·7 | 0·266 |
| Sugar | 23·0 | 23·0 | 0·000 |
| Short pastry | 241·0 | 175·5 | 2·310 |
| Mustard leaf protein (freeze-dried) | 81·0 | 78·6 | 9·400 |
| | 909·0 | 345·6 | 12·170 |

N as % of dry matter: calculated=3·52; by analysis=3·70. % of N due to lead protein= 77·3.

**Vegetable hot-pot with leaf protein**

*Method.* Parsnip, onion, swede, carrot, and turnip were diced and cooked with an oxo cube and water for 2 hr at 220° reducing to 120° for another 2 hr. A curry thickening was prepared by frying curry powder and spices, semolina and cooked moong dhal. Wheat leaf protein was added last, and the whole, now stiff, paste was fried a little longer. A known proportion of this paste was added to a known quantity of the vegetable hot-pot. The quantities can be altered to taste.

| Contents | Wt taken (g) | Dry wt (g) | N content (g) |
|---|---|---|---|
| (a) Parsnip | 100·0 | 17·50 | 0·270 |
| Onion | 99·0 | 7·12 | 0·148 |
| Swede | 108·5 | 9·34 | 0·195 |
| Carrot | 106·5 | 10·85 | 0·117 |
| Turnip | 118·0 | 7·91 | 0·142 |
| Oxo | 6·0 | 6·00 | 0·296 |
| Water | 692·0 | — | — |
| | 1230·0 | 58·72 | 1·168 |
| (b) Moong dahl | 40·5 | 40·5 | 1·860 |
| Semolina | 35·5 | 31·2 | 0·665 |
| Curry powder and spices | 62·0 | 62·0 | 0·943 |
| Corn oil | 34·5 | 34·5 | 0·000 |
| Wheat leaf protein (freeze-dried) | 66·5 | 62·9 | 6·800 |
| | 239·0 | 231·1 | 10·268 |
| (c) Vegetable hot-pot (a) | 342·5 | 34·25 | 0·681 |
| Indian panada (b) | 70·0 | 67·60 | 3·000 |
| | 412·5 | 101·85 | 3·681 |

N as % of dry matter: calculated=3·61; by analysis=3·29. % of N due to leaf protein= 54·0.

**Fish, cabbage and leaf protein pie**

*Method.* The haddock was poached in water, and then mixed with cooked finely shredded cabbage. Maize leaf protein and fish stock was added and the mixture put in a pie dish, covered with a thin pastry lid, and cooked at 230° for about 25 min.

This recipe shows how a conventional strongly flavoured protein source can be extended. In this dish about equal quantities of protein are contributed by the fish and the leaf protein. The pastry, which contributes approximately 10% of the total protein, could be omitted and the mixture served as a stew. Any other smoked or salted fish could be substituted for the haddock.

| Contents | Wt taken (g) | Dry wt (g) | N content (g) |
|---|---|---|---|
| Smoked haddock | 85·5 | 24·25 | 3·185 |
| Cabbage | 292·5 | 12·00 | 0·381 |
| Anchovy essence | 10·0 | 3·84 | 0·143 |
| Short pastry | 74·5 | 54·10 | 0·715 |
| Maize leaf protein (freeze-dried) | 30·0 | 29·50 | 2·830 |
| Water (fish stock) | 94·0 | — | — |
| | 586·5 | 123·69 | 7·254 |

N as % of dry matter: calculated = 5·87; by analysis = 5·64. % of N due to leaf protein = 39·0.

These dishes, as I have said, were planned to interest people in the U.K. Oke (unpublished) followed the same analytical regime in making some dishes suitable for use in Nigeria. Villagers there are used to taking maize gruel every morning with food mixtures ranging from black to dark green in colour, and so the colour of the leaf protein should not be a major problem. Hence a high supplementation was accepted with maize gruel especially by farmers. It seems the older they are the greater the supplementation they choose. Virtually all the food prepared for lunch and supper is taken with green vegetable stew which contains condiments such as crayfish, locust bean, pepper, tomatoes, etc. The product looks like a green mass of leaves. Sometimes palm oil is used (this depends on the vegetable) but usually it is omitted. In some cases the pepper soup is cooked separately and a little is added to the vegetable soup. Vegetable soups therefore afford a good medium for supplementation as long as the supplementation does not change the taste or colour. With certain foodstuffs, especially those that are coloured red by palm oil (e.g. jolloff rice), the green colour and gritty feel may create difficulties with respect to acceptability.

The dishes tried by Oke were:

Oyo, a typical green vegetable stew made by adding the chopped leaves to boiling water, stirring, and adding condiments and leaf protein. Its composition was:

| Contents | Wt taken (g) | Dry wt (g) | N content (g) |
|---|---|---|---|
| Vegetable, oyo (*Corchorus olitorius*) | 190·0 | 25·0 | 0·900 |
| Locust bean | 4·0 | 1·5 | 0·040 |
| Crayfish | 15·0 | 12·0 | 1·080 |
| Leaf protein | 24·0 | 23·6 | 2·400 |
| | 233·0 | 62·1 | 4·420 |

N as % of dry matter: calculated=7·20; by analysis=6·55. % of N due to leaf protein= 54·30.

Yam flour, a popular pudding in western Nigeria which is usually taken with oyo soup, is made by adding yam flour a little at a time to boiling water. When the mass is semi-solid, leaf protein is added and the mixture is stirred till homogeneous. The compositions of two batches were:

| Contents | Wt taken (g) | Dry wt (g) | N content (g) |
|---|---|---|---|
| (a) Yam flour | 60 | 48·0 | 0·72 |
| Water | 400 | — | — |
| Leaf protein | 10 | 10·0 | 1·00 |
| | 470 | 58·0 | 1·72 |

N as % of dry matter: calculated=2·97; by analysis=3·31. % of N due to leaf protein= 58·1.

| Contents | Wt taken (g) | Dry wt (g) | N content (g) |
|---|---|---|---|
| (b) Yam flour | 60 | 48·0 | 0·72 |
| Water | 400 | — | — |
| Leaf protein | 5 | 5·0 | 0·50 |
| | 465 | 53·0 | 1·22 |

N as % of dry matter: calculated=2·30; by analysis=1·98. % of N due to leaf protein= 41·0.

Maize gruel is a post-weaning diet in Nigeria. It is made by fermenting maize for about 3 days, filtering off the fine starch granules and making it into a paste with hot water. The coarse particles, consisting of the pericap, are used as animal feed. Thus maize gruel is virtually all carbohydrate. It is usually mixed with various other ingredients given by the native doctor for strength and good health. Leaf protein is added to the semi-solid flour paste and the mixture stirred till homogeneous. The compositions of two batches were:

| Contents | Wt taken (g) | Dry wt (g) | N content (g) |
|---|---|---|---|
| (a) Maize flour | 132 | 7·92 | 0·088 |
| Water | 525 | — | — |
| Leaf protein | 5 | 5·00 | 0·500 |
| | 662 | 12·92 | 0·588 |

N as % of dry matter: calculated=4·60; by analysis=4·65. % of N due to leaf protein= 85·0.

| Contents | Wt taken (g) | Dry wt (g) | N content (g) |
|---|---|---|---|
| (b) Maize flour | 79 | 4·74 | 0·053 |
| Water | 300 | — | — |
| Leaf protein | 5 | 5·00 | 0·500 |
| | 384 | 9·74 | 0·533 |

N as % of dry matter: calculated=5.70; by analysis=5·97. % of N due to leaf protein= 90·5.

The responses of various groups to whom these dishes were offered were interesting. They were acceptable to 8/8 farmers, 18/20 university staff, 9/10 school children but only 8/12 clerks. From the older villagers the only comment, even at the greater level of supplementation, was that the vegetables had been overcooked!

# Hilaire Marin Rouelle (1718–79)

H. M. Rouelle gets much less attention from the historians of chemistry than his more flamboyant elder brother Guillaume François Rouelle. Guillaume was an enthusiastic and outspoken teacher, his manner of lecturing was even described as bizarre. He taught many distinguished chemists (including Lavoisier) and was made demonstrator at the Jardin du Roi (now Jardin des Plantes) in Paris. His lectures were enlivened by such remarks as 'Écoutez-moi! car je suis le seul qui puisse vous démontrer ces vérités' and sometimes by explosions. In due course he was made professor and the rue Rouelle in Paris is named after him.

Hilaire had a very different character: neat, tactful and a skilled analyst. Guillaume's enthusiasm gradually turned into mania, he became irritable, petulant and, in the excitement of his lectures, sometimes partly undressed. In 1768 he became mentally unstable and handed over the lecture course to Hilaire who was made demonstrator in 1770. Hilaire was never given the title of professor but he has a distinguished chemical record. Having recognized the presence of potassium in cream of tartar, he made the tartrates of several other metals and showed that tartaric acid was present in unfermented grape juice. He isolated formic acid from ants, and separated urea, or possibly the hydrate of urea and sodium chloride, from urine. This substance had probably already been made in 1729 by Boerhaave. Besides these studies on substances with a definite composition, he studied various oils and resins and also the group of substances to which Berzelius was later to give the name 'protein'. He analysed blood and serous fluids, and repeated Beccari's observations on wheat gluten. Hitherto, starch and fat had been looked on as the nutritionally important components of foodstuffs; Rouelle stressed the importance of proteins. Beccari compared gluten to the animal proteins because they stank similarly when putrid. Rouelle commented on the similarity of their smells when heated. The title of his paper on leaf protein shows that he had a clear conception of proteins as a chemical category.

Internal evidence suggests that few of those who now quote Rouelle have had the opportunity of reading his paper and, until 1952, they were apt to give him either no initials or his brother's. Many of the observations that he made are, from time to time, rediscovered—they have even been patented. It seems worthwhile therefore to publish a free translation.

---

*Translation, by J. Leresche, of the paper 'Observations sur les Fécules ou parties vertes des Plantes, et sur la matière glutineuse ou végéto-animale' by Hilaire Marin Rouelle (1773).*

## Observations on the sediments or green parts of plants and on glutinous or vegeto-animal matter; by M. Rouelle, demonstrator in chemistry in the 'Jardin du Roi'

The sediments or green parts of plants were classed as resins by my late brother because of their solubility in all oily solvents and in alcohol. He defined these sediments as being composed of (1) a part of green resinous colouring matter and (2) a part of parenchyma or plant fibres, separated by pounding with a pestle; and he pointed out that when one heated either the sediments or the plant juices with the oils or fats that dissolve the green part, there always remained some insoluble matter which he considered, as I have just said, to belong to the parenchyma or fibrous portion of the plant.

I have given, in the *Journal de Médecine* of last March and in the *Avant Coureur*, etc., the analyses of several sediments or green parts of plants. I have explained that sediments made from different plant families, after having been dried, give, when analysed using a retort, the same products as animal substances; which proves that the sediments or green coloured parts of plants are not made up of pure vegetable matter, because the products from the analysis of vegetable matter are not found in them, but on the contrary, those of animal matter.

When I wrote of this first work in the *Journal de Médecine* and when I said that 'the green sediments were not resins as the products of their analysis were quite different from those of all known resins', I thought I did not need to give a clearer explanation of the nature of this substance; I stated, however, that one could demonstrate the presence in all vegetable matter of a substance completely similar to the glutinous matter of wheat. I laid aside for a further paper, the more precise description of this kind of substance which is in fact composed of a pure resin which gives the green colour to all plants and of this glutinous or vegeto-animal matter.

The glutinous matter, which is found in all the sediments, is generally there in more abundance than the resinous green part, the latter making up, in the plants that I have examined, only a third, fourth and fifth part of the sediment; I will confine myself for the moment to a single example and I will indicate some others.

**Sediment of hemlock**

One takes the required quantity of hemlock when it is almost in flower. One grinds it carefully in a marble mortar with a wooden pestle. One presses the plant, and strains the juice that one obtains through a well-stretched cloth or a cloth filter. One heats this juice to the point at which one can hold one's finger in it for several minutes. The sediment separates out and part of it floats on top of the liquid, another part of it precipitates out or remains floating in the juice. The whole is passed through a cloth filter, the liquid becomes clear, and the sediment remains on the cloth and is carefully collected. This is the procedure that one normally follows to obtain these plant sediments.

**Remarks**

I. If one pays a little attention to what is going on in this process, one sees that the part of the sediment which separates out first is the greenest part. By increasing the heat, one can see very distinctly underneath the part of the sediment which floats on the liquid, a portion of the substance which forms little flakes or white dots which are very distinct and well defined; these observations suggest that there are two sediments.

II. If one only heats the juice until it has the heat of milk when it is drawn from a cow and if one takes the vessel off the fire, the sediment which separates out has a more beautiful green colour than the one in Remark I. If one immediately pours the juice onto a cloth, the liquid which comes through is still coloured with a slight tinge of green.

III. If this liquid which has been separated as above from the green part, is heated more strongly than at first, a sediment still separates out which is slightly green-coloured and a dirty white. This second sediment contains a much greater amount of glutinous, or vegeto-animal matter, than the first. It does however contain an abundance of it as will be seen.

IV. The sediment obtained by the normal method is put in earthenware or glazed vessels. It is carefully diluted using a wooden stirrer, by pouring in little by little up to 8 or 9 *pintes* of water per vessel. One leaves it all to stand for 24 hr, so as to give the sediment time to precipitate out. One decants the water and pours on an equal quantity a second and a third time, leaving it to stand 24 hr again each time. After the third washing, one puts the sediment on to a cloth mounted on a wooden frame, so as to get as much moisture out of it as possible; then one puts the sediment with the cloth on to a plaster tile or slab so that most of the water that it still contains will be absorbed: the sediment then becomes fairly solid; one cuts it into little pieces and puts it to dry on pieces of paper placed on sieves.

V. One could make use of several methods to separate the two sediments; but some of them are not easily practicable, and others are too expensive. The easiest is the method that uses alcohol which, as one knows, has no action on vegeto-animal matter, whereas on the contrary it dissolves the green colouring matter. One has only to grind the dried sediment to a fine powder and to digest it with alcohol several times.

The alcohol dissolves the green part and leaves the vegeto-animal part; but the process takes a very long time because of the way the two substances adhere during drying; I have even observed that it was very difficult to separate them completely.

VI. The two substances are more easily separated using the fresh sediment if one follows the above procedure (Remark IV). When it is ready to be dried out, one puts it in a marble mortar and, using a glass or wooden pestle, one carefully and slowly dilutes it with alcohol. One pours it all into a glass vessel big enough to hold 8–9 *pintes* of alcohol. One digests it for 24 hr in a water-bath or on a sand-bath; one leaves it to cool and then decants off the alcohol which one filters. Eight or 9 *pintes* of fresh alcohol are poured on again and the digestion is repeated for a second and a third time in the same way as the first; and after having obtained the three extracts, one places them in the tin boiler of a distilling apparatus, so as to distill off all the alcohol which passes over as a clear liquid which has a smell of hemlock. There remains in the boiler a resinous substance which is very soft and which clings to the fingers like terebinth resin. The alcohol that has been distilled off is poured back successively in three portions on to the sediment which is in the glass vessel where it still takes up a strong green colour. This new tincture is distilled like the first one and the distillation as well as the digestions of this same alcohol are repeated until the sediment no longer gives it any colour, and it comes out clear. The quantity of green colouring matter that one obtains from such an amount of sediment varies a little; it depends firstly on the state and age of the plant and secondly on the part of the plant that one takes, the leaves or the stalks, or the whole plant.

VII. The sediment which remains in the glass vessel has a dirty greyish-white colour, and blackens after being dried. It alone makes up three-quarters of the quantity of matter used in the experiment; it is this that is the glutinous or vegeto-animal substance, as its analysis will show.

VIII. *Analysis of the glutinous or vegeto-animal substance, separated from the green colouring matter of the sediment of hemlock.* I placed in the reverberatory furnace a retort in which I had put 4 *onces* of this substance. When it was slowly heated, a small amount of distillate came over first, then the first drops that came after it were volatile alkali. By gradually increasing the heat, this volatile alkali was concentrated and at the end of the distillation I obtained solid alkali. At the same time an oil that floated on the alkaline fluid passed over, this was like the one that one obtains from the vegeto-animal matter of flour and from the caseous part of milk.

The residue is fairly bulky. The pieces of the glutinous substance softened and collected together, so that the residue is of a fairly even texture and very much resembles the residue obtained from the glutinous part of corn and from the caseous part of milk. It weighs an *once* or more.

IX. *Distillation of the green part, or sediment of hemlock, separated by alcohol from the vegeto-animal matter.* I put 2 *onces* of the green colouring matter in a glass retort. This retort was placed in the reverberatory furnace over a naked flame with a glass flask as a receiver. I heated it little by little and a few drops of distillate passed over. When the heat was increased, an acid liquid came over which got stronger as it continued to come over and a clear oil coloured a beautiful yellow which became darker as it thickened. The acid that comes off is very strong and is similar to that obtained from wax. The oil is light and floats on the acid, as most of the oils obtained from resins do. Finally the residue is fairly bulky and light and weighs 3 *gros* and 60 *grains*.

X. Rosemary, from which the extractable part has been removed by repeated decoctions, is only, according to Boerhaave, the dross or skeleton of the plant. It still contains a small amount of crude oil which gives a little flame, but its ash does not give fixed alkali.

This learned doctor did not know that this exhausted rosemary gives fixed alkali and contains a green colouring matter which is soluble in fats, oils, resins and alcohol.

My brother, as I have already said, and as we have written, it was indeed printed long ago, and is in short known by everybody, was the first to have demonstrated this green part in exhausted rosemary; but as well as that, there are also in this rosemary, left after extraction with water and alcohol, two substances, namely: (1) a very small quantity of glutinous or vegeto-animal matter and (2) a substance that has a vegetable nature, insoluble in water and alcohol. The rosemary left after extraction with these two solvents, when distilled, still gives to a fairly marked extent some acid and some oil, consequently there remains in the rosemary a constituent or a substance which was hitherto unknown and which escapes the action of these two solvents. Later, I will describe this substance more fully.

---

Rouelle seems to use the word *fécule* in more than one sense. In modern French usage it means starch, but in older usage the word was applied to any sediment. Old weights and measures are imprecise; *onces, pintes, gros* and *grains* have therefore been left in the original form.

# References

AHMED S.Y. & SINGH N. (1969). A heat coagulation unit for leaf protein. *Indian J. Technol* **7**, 411.

AKESON W.R. & STAHMANN M.A. (1965). Nutritive value of leaf protein concentrate, an *in vitro* digestion study. *J. agric. Fd Chem.* **13**, 145.

AMERINE M.A., PANGBORN R.M. & ROESSLER E.B. (1965). *Principles of Sensory Evaluation of Foods*, p. 276. Academic Press, New York.

ARKCOLL D.B. (1969). Preservation of leaf protein preparations by air drying. *J. Sci. Fd Agric.* **20**, 600.

ARKCOLL D.B. & FESTENSTEIN G.N. (1971). A preliminary study of the agronomic factors affecting the yields of extractable leaf protein. *J. Sci. Fd Agric.* **22**, 49.

ARKCOLL D.B. & HOLDEN M. (1971). Carotenoids in leaves and leaf protein. *Rep. Roth. exp. Stn for* 1970 (in press).

ARONOFF S. (1953). The chemistry of chlorophyll, with especial reference to foods. *Adv. Fd Res.* **4**, 133.

BARBER R.S., BRAUDE R. & MITCHELL K.G. (1959). Leaf protein in rations of growing pigs. *Proc. Nutr. Soc.* **18**, iii.

BENDER A.E. (1956). Relation between protein efficiency and net protein utilization. *Brit. J. Nutr.* **10**, 135.

BICKOFF E.M., BEVENUE A. & WILLIAMS K.T. (1947). Alfalfa has a promising chemurgic future. *Chemurgic Digest*, **6**, 213.

BIDMEAD D.S. & LEY F.J. (1958). Quantitative amino acid analysis of food proteins by means of a single ion-exchange column. *Biochim. biophys. Acta*, **29**, 562.

BLANKHART D.M. (1967). Levels of carotene in diet of Indonesian children. *Trop. geogr. Med.* **19**, 144.

BLOCK R.J. & MITCHELL, H.H. (1946). The correlation of the amino acid composition of proteins with their nutritive value. *Nutr. Abstr. Rev.* **16**, 249.

BOULTER D. (1966). An introduction to automatic amino acid analysis with plant extracts. *Techniques in Amino Acid Analysis*, p. 93. Technicon Instruments Co. Ltd, Chertsey, Surrey.

BOYD C.E. (1968). Fresh-water plants: a potential source of protein. *Econ. Bot.* **22**, 359.

BOYD C.E. (1969a). The nutritive value of three species of water weeds. *Econ. Bot.* **23**, 123.

BOYD C.E. (1969b). Production, mineral nutrient absorption, and biochemical assimilation by *Justicia americana* and *Alternanthera philoxeroides*. *Arch. Hydrobiol.* **66**, 139.

BOYD C.E. (1970a). Vascular aquatic plants for mineral nutrient removal from polluted waters. *Econ. Bot.* **24**, 95.

BOYD C.E. (1970b). Amino acid, protein and caloric content of vascular aquatic macrophytes. *Ecology,* **51,** 902.

BOYD C.E. & HESS L.W. (1970). Factors influencing shoot production and mineral nutrient levels in *Typha latifolia. Ecology,* **51,** 296.

BRYANT M. & FOWDEN L. (1959). Protein composition in relation to age of daffodil leaves. *Ann. Bot.* N.S. **23,** 65.

BUCHANAN R.A. (1969a). *In vivo* and *in vitro* methods of measuring nutritive value of leaf protein preparations. *Br. J. Nutr.* **23,** 533.

BUCHANAN R.A. (1969b). Effect of storage and lipid extraction on the properties of leaf protein. *J. Sci. Fd Agric.* **20,** 359.

BUJARD E., HANDWERCK V. & MAURON J. (1967). The differential determination of lysine in heated milk. I. *In vitro* methods. *J. Sci. Fd Agric.* **18,** 52.

BYERS M. (1961). The extraction of protein from the leaves of some plants growing in Ghana. *J. Sci. Fd Agric.* **12,** 20.

BYERS M. (1965). *Rep. Rothamsted exp. Stn for* 1964, p. 121.

BYERS M. (1967). The *in vitro* hydrolysis of leaf proteins. I. The action of papain on protein concentrates extracted from the leaves of *Zea mays. J. Sci. Fd Agric.* **18,** 28.

BYERS M. (1970a). *Rep. Rothamsted exp. Stn for* 1969 (*Part I*), p. 133.

BYERS M. (1970b). *Rep. Rothamsted exp. Stn for* 1969 (*Part I*), p. 134.

BYERS M. (1971). The amino acid composition and *in vitro* digestibility of some protein fractions from three species of leaves of various ages. *J. Sci. Fd Agric.* (in press).

BYERS M. & DAVYS M.N.G. (1964). *Rep. Rothamsted exp. Stn for* 1963, p. 96.

BYERS M., GREEN S.H. & PIRIE N.W. (1965). The presentation of leaf protein on the table II. *Nutrition,* **19,** 63.

BYERS M. & JENKINS G. (1961). Effect of gibberellic acid on the extraction of protein from the leaves of spring vetches (*Vicia sativa L.*). *J. Sci. Fd Agric.* **12,** 656.

BYERS M. & STURROCK J.W. (1965). The yields of leaf protein extracted by large-scale processing of various crops. *J. Sci. Fd Agric.* **16,** 341.

CARPENTER K.J. (1960). The estimation of the available lysine in animal-protein foods. *Biochem. J.* **77,** 604.

CARPENTER K.J., DUCKWORTH J. & ELLINGER G.M. (1952). The supplementary protein value of a by-product from grass processing. *Br. J. Nutr.* **6,** xii.

CARPENTER K.J., DUCKWORTH J. & ELLINGER G.M. (1954). The value of herbage concentrates for non-ruminants. *Proc. European Grassland Conf., Paris,* p. 243.

CASSELMAN T.W., GREEN V.E., ALLEN R.J. & THOMAS F.H. (1965) Mechanical dewatering of forage crops. *Tech. Bull.* 694, Agric. exp. Stn, Univ. Florida, Gainesville.

CHAYEN I.H., SMITH R.H., TRISTRAM G.R., THIRKELL D. & WEBB T. (1961). The isolation of leaf components. I. *J. Sci. Fd Agric.* **12,** 502.

CHIBNALL A.C. (1939). *Protein Metabolism in the Plant.* Yale University Press, New Haven.

CHIBNALL A.C., REES M.W. & LUGG J.W.H. (1963). The amino acid composition of leaf proteins. *J. Sci. Fd Agric.* **14,** 234.

CHIBNALL A.C. & SCHRYVER S.B. (1921). Investigations on the nitrogenous metabolism of the higher plants. I. The isolation of proteins from leaves. *Biochem. J.* **15,** 60.

CHRISMAN J. & KOHLER G.O. (1968). Separation milling of alfalfa. *Tenth Tech. Alf. Conf. Proc* ARS.-74-46, p. 37.

COONEY W.T., BUTTS J.S. & BACON L.E. (1948). Alfalfa meal in chick rations. *Poult. Sci.* **27,** 828.

COWLISHAW S.J. & EYLES D.E. (1956). The value of leaf protein concentrates for growing and laying pullets. *Emp. J. exp. Agric.* **24,** 223.

COWLISHAW S.J., EYLES D.E., RAYMOND W.F. & TILLEY J.M.A. (1956). Nutritive value of leaf protein concentrates. I. Effect of addition of cholesterol and amino acids. *J. Sci Fd Agric.* **7,** 768.

CRESTFIELD A.M., MOORE S. & STEIN W.H. (1963). The preparation and enzymatic hydrolysis of reduced and S-carboxymethylated proteins. *J. biol. Chem.* **238,** 622.

CROOK E.M. (1946). The extraction of nitrogenous materials from green leaves. *Biochem. J.* **40,** 197.

DATTA R.K., CHAKRABARTY P.R., GUHA B.C. & GHOSH J.J. (1966). Protein concentrates from leaves of water hyacinth. *Indian J. appl. Chem.* **29,** 7.

DAVIES M., EVANS W.C. & PARR W.H. (1952). Biological values and digestibilities of some grasses, and protein preparations from young and mature species, by the Thomas-Mitchell method, using rats. *Biochem. J.* **52,** xxiii.

DAVYS M.N.G. & PIRIE N.W. (1960). Protein from leaves by bulk extraction. *Engineering,* **190,** 274.

DAVYS M.N.G. & PIRIE N.W. (1963). Batch production of protein from leaves. *J. agric. Engng Res.* **8,** 70.

DAVYS M.N.G. & PIRIE N.W. (1965). A belt press for separating juices from fibrous pulps. *J. agric. Engng Res.* **10,** 142.

DAVYS M.N.G. & PIRIE N.W. (1969). A laboratory-scale pulper for leafy plant material. *Biotech. Bioengng,* **11,** 517.

DAVYS M.N.G., PIRIE N.W. & STREET G. (1969). A laboratory-scale press for extracting juice from leaf pulp. *Biotech. Bioengng,* **11,** 528.

DEVADAS R.P. (1967). Acceptability of novel proteins. *J. Nutr. Dietet.* **6,** 26.

DEVADAS R.P. & EASWARAN P.P. (1967). Influence of socio-economic factors on the nutritional status and food intake of preschool children in a rural community. *J. Nutr. Dietet.* **4,** 155.

DEVADAS R.P., LYZAMMA M. & RADHA RUKMANI A. (1970). Evaluation of the supplementary value of the leaf protein concentrates using albino rats. *Indian J. Nutr. Dietet.* **7,** 234.

DORAISWAMY T.R., SINGH N. & DANIEL V.A. (1969). Effects of supplementing ragi (*Eleusine coracana*) diets with lysine or leaf protein on the growth and nitrogen metabolism of children in India. *Br. J. Nutr.* **23,** 737.

DUCKWORTH J., HEPBURN W.R. & WOODHAM A.A. (1961). Leaf protein concentrates. II. The value of a commercially dried product for newly-weaned pigs. *J. Sci. Fd Agric.* **12,** 16.

DUCKWORTH J. & WOODHAM A.A. (1961). Leaf protein concentrates. I. Effect of source of raw material and method of drying on protein value for chicks and rats. *J. Sci. Fd Agric.* **12,** 5.

DUCKWORTH J., WOODHAM A.A. & MCDONALD I. (1961). The assessment of nutritive value in protein concentrates by the Gross Protein Value method. *J. Sci. Fd Agric.* **12,** 407.

DUSTIN J.P., CZAJKOWSKA C., MOORE S. & BIGWOOD E.J. (1953). A study of the chromatographic determination of amino acids in the presence of large amounts of carbohydrate. *Analytica chim. Acta,* **9,** 256.

EAKER D.D. (1968). In the *full text* of 'The determination of free and protein-bound amino acids', given at the symposium on 'Evaluation of Novel Protein Products', Stockholm, Sweden, 1968. [These details omitted in version published 1970 (Pergamon Press, Oxford)].

EGGUM B.O. (1970a). Nutritional evaluation of proteins by laboratory animals. *Evaluation of Novel Protein Products* (Ed. by A.E. Bender *et al*), p. 117. Pergamon Press, Oxford.

EGGUM B.O. (1970b). The protein quality of cassava leaves. *Br. J. Nutr.* **24,** 761.

ELLINGER G.M. (1954). The evaluation of leaf protein concentrates in poultry rations. *Proc. World's Poult. Congr.* (No. 10, Edinburgh), **2,** 128.

ELLINGER G.M. & BOYNE E.B. (1965). Amino acid composition of some fish products and casein. *Br. J. Nutr.* **19,** 587.

ELLINGER G.M. & PALMER R. (1969). The biological availability of methionine sulphoxide. *Proc. Nutr. Soc.* **28,** 42A.

EREKY K. (1926). Vegetable foods and medicines for men and animals. *Brit. Pat.* 270629.

F.A.O./W.H.O. (1965). Protein requirements. *F.A.O. Nutr. Mtg. Rep. Ser.* no. 37 (Food and Agriculture Organization, Rome).

F.A.O. (1968). *Production Yearbook,* **22.**

FEENY P.P. (1969). Inhibitory effect of oak leaf tannins on the hydrolysis of proteins by trypsin. *Phytochemistry,* **8,** 2119.

FORD J.E. (1962). A microbiological method of assessing the nutritional value of proteins. 2. The measurement of 'available' methionine, leucine, iso-leucine, arginine, histidine, tryptophan and valine. *Br. J. Nutr.* **16,** 409.

FORD J.E. (1964). A microbiological method of assessing the nutritional value of proteins. 3. Further studies on the measurements of available amino acids. *Br. J. Nutr.* **18,** 449.

FORD J.E. (1970). Personal communication.

GERLOFF E.D., LIMA I.H. & STAHMANN M.A. (1965). Amino acid composition of leaf protein concentrates. *J. agric. Fd Chem.* **13,** 139.

GLENCROSS R.G. (1969). Unpublished results.

GOLDSTEIN J.L. & SWAIN T. (1965). The inhibition of enzymes by tannins. *Phytochemistry,* **4,** 185.

GOODALL C. (1936). Improvements relating to the treatment of grass and other vegetable substances. *Brit. Pat.* 457789.

GORDON A.J. & TOPPS J.H. (1970). Unpublished results.

HARTLEY H. (1951). Origin of the word 'protein'. *Nature, Lond.* **168,** 244.

HARTMAN G.H., AKESON W.R. & STAHMANN M.A. (1967). Leaf protein concentrate prepared by spray-drying. *J. agric. Fd Chem.* **15,** 74.

HARVEY D. (1970). *Tables of the Amino Acids in Foods and Feedingstuffs,* Commonwealth Bureau of Animal Nutrition, Technical Communication No. 19, 2nd Edition, p. 95 (Commonwealth Agricultural Bureau, Farnham Royal, Bucks.).

HASLER A.D. (1969). Cultural eutrophication is reversible. *Bioscience,* **19,** 425.

HAWKINS G.E. (1959). Relationships between chemical composition and some nutritive qualities of *Lespedeza serica* hays. *J. Anim. Sci.* **18,** 763.

HEIMAN V., CARVER J.S. & COOK, J.W. (1939). A method of determining the gross value of protein concentrates. *Poult. Sci.* **18,** 464.

HENRY K.M. (1963). The nutritive value of leaf proteins. *Proc. 6th Int. Nutr. Congr., Edinburgh,* p. 492.

HENRY K.M. & FORD J.E. (1965). The nutritive value of leaf protein concentrates determined in biological tests with rats and by microbiological methods. *J. Sci. Fd Agric.* **16,** 425.

HOLM L.G., WELDON L.W. & BLACKBURN R.D. (1969). Aquatic weeds. *Science,* **166,** 699.

HORIGOME T. & KANDATSU M. (1968). Biological value of proteins allowed to react with phenolic compounds in presence of *o*-diphenol oxidase. *Agr. biol. Chem.* **32,** 1093.

HUGHES G.P. & EYLES D.E. (1953a). Extracted herbage leaf proteins for poultry feeding. I. Introduction and feeding trial with laying hens. *J. agric. Sci.* **43,** 136.

HUGHES G.P. & EYLES D.E. (1953b). Extracted herbage leaf proteins for poultry feeding. II. The use of leaf protein in chick rations. *J. agric. Sci.* **43,** 144.

HUGHES G.P. & EYLES D.E. (1953c) Extracted herbage leaf proteins for poultry feeding. III. The digestibility of the residual product resulting from the extraction of herbage leaf protein. *J. agric. Sci.* **43,** 152.

HUME E.M. & KREBS H.A. (1949). *Vitamin A Requirement of Human Adults.* Medical Research Council, Special Report Series No. 264.

IKAWA M. & SNELL E.E. (1961). Artifact production through esterification of glutamic acid during analytical procedures. *J. biol. Chem.* **236,** 1955.

INDIAN COUNCIL OF MEDICAL RESEARCH (1968). Recommended daily allowances of nutrient and balanced diets. *Nutrition Research Laboratories,* p. 2.

INDIAN COUNCIL OF MEDICAL RESEARCH (1969). Annual report of National Institute of Nutrition, p. 68.

JENNINGS A.C., PUSZTAI A., SYNGE R.L.M. & WATT W.B. (1968). Fractionation of plant material. III. Two schemes for chemical fractionation of fresh leaves having special applicability for isolation of the bulk protein. *J. Sci. Fd Agric.* **19,** 203.

JÖNSSON A.G. (1962). Studies in the utilization of some agricultural wastes and by-products by various microbial processes. *K. Lantbrhögsk. Annlr.* **28,** 235.

KAMALANATHAN G., NALINAKSHI G.S. & DEVADAS R.P. (1970). Effect of a blend of protein foods on the nutritional status of pre-school children in a rural blawadi. *Indian J. Nutr. Dietet.* **7,** 288.

KAMALANATHAN G., USHA M.S. & DEVADAS R.P. (1969). Evaluation of acceptability of some recipes with leaf protein concentrates. *J. Nutr. Dietet.* **6,** 12.

KIESEL A., BELOZERSKY A., AGATOV P., BIWSCHICH N. & PAWLOWA M. (1934). Vergleichende Untersuchungen über Organeiweiss von Pflanzen. *Z. physiol. Chem.* **226,** 73.

KOHLER G.O., BICKOFF E.M., SPENCER R.R., WITT S.C. & KNUCKLES B.E. (1968). Wet processing of alfalfa for animal feed products. *Tenth Tech. Alf. Conf. Proc.* ARS–74–46, 71–79.

KOHLER G.O. & GRAHAM W.R., JR (1951). A chick growth factor found in leafy green vegetation. *Poult. Sci.* **30,** 484.

KUPPUSWAMY S., SRINIVASAN M. & SUBRAHMANYAN V. (1958). *Protein in Foods,* p. 221 (Ind. Council Med. Res., New Delhi).

LALA V.R. & REDDY V. (1970). Absorption of β-carotene from green leafy vegetables in undernourished children. *Am. J. clin. Nutr.* **23,** 110.

LEXANDER K., CARLSSON R., SCHALEN V., SIMONSSON A. & LUNDBORG T. (1970). Quantities and qualities of leaf protein concentrates from wild species and crop species grown under controlled conditions. *Ann. appl. Biol.* **66,** 193.

LIMA I.H., RICHARDSON T. & STAHMANN M.A. (1965). Fatty acids in some leaf protein concentrates. *J. agric. Fd Chem.* **13,** 143.

LITTLE E.C.S. (1968). *Handbook of Utilization of Aquatic Plants.* F.A.O., Rome.

LIVERMORE D.F. & WUNDERLICH W.E. (1969). Mechanical removal of organic production from waterways. *Eutrophication: Causes, Consequences, Correctives,* p. 494. National Academy of Sciences, Washington.

LOUGH A.K. (1968). Fatty acid compositions of fractions of broad-bean leaves. *Biochem. J.* **107,** 28P.

LUGG J.W.H. (1939) The representativeness of extracted samples and the efficiency of extraction of protein from the fresh leaves of plants; and some partial analyses of the whole proteins of leaves. *Biochem. J.* **33A,** 110.

LUGG J.W.H. (1946). Problems associated with the acid hydrolysis of an impure protein preparation. *Biochem. J.* **40,** 88.

LUGG J.W.H. & WELLER R.A. (1948). Protein in senescent leaves of *Trifolium subterraneum:* partial amino acid composition. *Biochem. J.* **43,** 412.

MAHADEVIAH S. & SINGH N. (1968). Leaf protein from the green tops of *Cichorium intybus* L. (chicory). *Indian J. exp. Biol.* **6,** 193.

MCNAUGHTON S.J. (1966). Ecotype function in the *Typha* community-type. *Ecol. Monogr.* **36,** 297.

MILLER D.S. (1965). Some nutritional problems in the utilization of non-conventional proteins for human feeding. *Recent Adv. Fd Sci.* **3,** 125.

MILLER D.S. & DONOSO G. (1963). Relationship between the sulphur/nitrogen ratio and the protein value of diets. *J. Sci. Fd Agric.* **14,** 345.

MILLER D.S. & NAISMITH D.J. (1959). The estimation of net dietary-protein value (N.D.-p.V.) of meals and diets from the total sulphur content. *Proc. Nutr. Soc.* **18,** viii.

MILLER D.S. & SAMUEL P. (1968). Methionine sparing compounds. *Proc. Nutr. Soc.* **27,** 21A.

MILLER E.L. (1967). Determination of the tryptophan content of feedingstuffs with particular reference to cereals. *J. Sci. Fd Agric.* **18,** 381.

MILLER E.L., HARTLEY A.W. & THOMAS D.C. (1965). Availability of sulphur amino acids in protein foods. 4. Effect of heat treatment upon the total amino acid content of cod muscle. *Br. J. Nutr.* **19,** 565.

MORRISON J.E. & PIRIE N.W. (1961). The large scale production of protein from leaf extracts. *J. Sci. Fd Agric.* **12,** 1.

NAZIR M. & SHAH F.H. (1966). Extractability of proteins from various leaves. *Pakist. J. scient. ind. Res.* **9,** 235.

NOWAKOWSKI T.Z. & CUNNINGHAM R.K. (1966). Nitrogen fractions and soluble carbohydrates in Italian Ryegrass. II. Effects of light intensity, form and level of nitrogen. *J. Sci. Fd Agric.* **17,** 145.

OELSHLEGEL F.J., JR, SCHROEDER J.R. & STAHMANN M.A. (1969a). Potential for protein concentrates from alfalfa and waste green plant material. *J. agric. Fd Chem.* **17,** 791.

OELSHLEGEL F.J., JR, SCHROEDER J.R. & STAHMANN M.A. (1969b). Protein concentrates: Use of residues as silage. *J. agric. Fd Chem.* **17,** 796.

OOMEN H.A.P.C. (1961). An outline of xerophthalmia. *Int. Rev. trop. Med.* **1,** 131.

OSBORNE T.B. & WAKEMAN A.J. (1920). The proteins of green leaves. *J. biol. Chem.* **42,** 1.

OSTROWSKI H., JONES A.S. & CADENHEAD A. (1970). Availability of lysine in protein concentrates and diets using Carpenter's method and a modified Silcock method. *J. Sci. Fd Agric.* **21,** 103.

PELLETT P.L., KANTARJIAN A. & JAMALIAN J. (1969). Use of sulphur analysis in the production of the protein value of middle-eastern diets. *J. Sci. Fd Agric.* **20,** 229.

PENFOUND W.T. (1956). Primary production of vascular aquatic plants. *Limnol. Oceanogr.* **1,** 92.

PENFOUND W.T. & EARLE T.T. (1948). The biology of the water hyacinth. *Ecol. Monogr.* **18,** 448.

PETERS J.P. & VAN SLYKE D.D. (1932). *Quantitative Clinical Methods,* Vol. II, p. 395. Bailliere, Tindall & Cox, London.

PETERSON D.W. (1950). Some properties of a factor in alfalfa meal causing depression of growth in chicks. *J. biol. Chem.* **183,** 647.

PIERPOINT W.S. (1969a). *o*-Quinones formed in plant extracts. Their reactions with amino acid and peptides. *Biochem. J.* **112,** 609.

PIERPOINT W.S. (1969b). *o*-Quinones formed in plant extracts. Their reaction with bovine serum albumin. *Biochem. J.* **112,** 619.

PIERPOINT W.S. (1971). Formation and behaviour of *o*-quinones in some processes of agricultural importance. *Rep. Rothamsted exp. Stn for* 1970, p. 199.

PIRIE N.W. (1942). Green leaves as a source of proteins and other nutrients. *Nature, Lond.* **149,** 251.

PIRIE N.W. (1950). The isolation from normal tobacco leaves of nucleo-protein with some similarity to plant viruses. *Biochem. J.* **47,** 614.

PIRIE N.W. (1953). Large-scale production of edible protein from fresh leaves. *Rep. Rothamsted exp. Stn for* 1952, p. 173.

PIRIE N.W. (1957). Macromolecular nucleoproteins from healthy tobacco leaves. *Biochimia* **22,** 140.

PIRIE N.W. (1959a). Edible protein from green leaves. *Br. Vegetarian,* **1,** 229.

PIRIE N.W. (1959b). The large-scale separation of fluids from fibrous pulps. *J. biochem. microbiol. Technol. Engng,* **1,** 13.

PIRIE N.W. (1961). The disintegration of soft tissues in the absence of air. *J. agric. Engng Res.* **6,** 142.

PIRIE N.W. (1964). Freeze drying, or drying by sublimation. *Instrumental Methods of Experimental Biology* (Ed. by D.W. Newman), p. 189. Macmillan, New York.

PIRIE N.W. (1966a). Leaf protein as a human food. *Science, N.Y.* **152,** 1701.

PIRIE N.W. (1966b). Fodder fractionation: an aspect of conservation. *Fertil. Feed. Stuffs J.* **63,** 119.

PIRIE N.W. (1968a). *Rep. Rothamsted exp. Stn for* 1967, p. 110.

PIRIE N.W. (1968b). Food from the forests. *New Scient.* **40,** 420.

PIRIE N.W. (1969a). The production and use of leaf protein. *Proc. Nutr. Soc.* **28,** 85.

PIRIE N.W. (1969b). The present position of research on the use of leaf protein as a human food. (International symposium on protein foods and concentrates, Mysore (India) 1967). *Pl Foods Human Nutr.* **1,** 237.

PIRIE N.W. (1969c). *Food Resources: 'Conventional and Novel'*. Penguin Books, London.

PIRIE N.W. (1970a). Leaf proteins. *Symposium 'Evaluation of Novel Protein Products'*, p. 87. Pergamon Press, Oxford.

PIRIE N.W. (1970b). Weeds are not all bad. (Fresh-water weeds as a resource). *Ceres*, **3**, 31.

PLESHKOV B.P. & FOWDEN L. (1959). Amino acid composition of the proteins of barley leaves in relation to the mineral nutrition and age of plants. *Nature, Lond.* **183**, 1445.

PORTER J.W.G., WESTGARTH D.R. & WILLIAMS A.P. (1968). A collaborative test of ionexchange chromatographic methods for determining amino acids. *Br. J. Nutr.* **22**, 437.

RAO S.R., CARTER F.L. & FRAMPTON V.L. (1963). Determination of available lysine in oilseed meal proteins. *Analyt. Chem.* **35**, 1927.

RAO C.N. & RAO B.S. (1970). Absorption of dietary carotenes in human subjects. *Am. J. clin. Nutr.* **23**, 105.

RAYMOND W.F. & HARRIS C.E. (1957). The value of the fibrous residue from leaf protein extraction as a feeding stuff for ruminants. *J. Br. Grassld Soc.* **12**, 166.

RAYMOND W.F. & TILLEY J.M.A. (1956). The extraction of protein concentrates from leaves. *Colon. Pl. Anim. Prod.* **6**, 3.

ROACH A.G. (1966). The preparation of feedstuffs and samples for amino acid analysis. *Techniques in Amino Acid Analysis*, p. 86. Technicon Instruments Co. Ltd, Chertsey, Surrey.

ROACH A.G., SANDERSON P. & WILLIAMS D.R. (1967). Comparison of methods for the determination of available lysine value in animal and vegetable protein sources. *J. Sci. Fd Agric.* **18**, 274.

ROACH D. & GEHRKE C.W. (1970). The hydrolysis of proteins. *J. Chromat.* **52**, 393.

ROUELLE H.M. (1773). Observations sur les fécules ou parties vertes des plantes, et sur la matière glutineuse ou végéto-animale. *J. de Médecine, Chirurgie, Pharmacie, etc.* **40**, 59.

SENTHESHANMUGANATHAN S. & DURAND S. (1969). Isolation and composition of proteins from leaves of plants grown in Ceylon. *J. Sci. Fd Agric.* **20**, 603.

SHAH F.H., RIAZ-UD-DIN & SALAM A. (1967). Effect of heat on the digestibility of leaf proteins. I. Toxicity of the lipids and their oxidisation products. *Pakist. J. scient. ind. Res.* **10**, 39.

SHURPALEKAR K.S., SINGH N. & SUNDARAVALLI O.E. (1969). Nutritive value of leaf protein from lucerne (*Medicago sativa*): Growth responses in rats at different protein levels and to supplementation with lysine and/or methionine. *Indian J. exp. Biol.* **7**, 279.

SINGH N. (1960). Differences in the nature of nitrogen precipitated by various methods from wheat leaf extracts. *Biochim. biophys. Acta*, **45**, 422.

SINGH N. (1962). Proteolytic activity of leaf extracts. *J. Sci. Fd Agric.* **13**, 325.

SINGH N. (1964). Leaf protein extraction from some plants of Northern India. *J. Fd Sci. Tech.* **1**, 37.

SINGH N. (1968). Leaf proteins in nutrition—Studies on production, nutritive value and utilisation. *Voeding*, **30**, 710.

SINGH N. (1967, 1969, 1970). In Central Food Technological Research Institute (Mysore) reports.

SLADE R.E. (1937). Grass and the national food supply. *Rep. Br. Ass. Advmt Sci.* 457.

SLADE R.E. & BIRKINSHAW J.H. (1939). Improvements in or relating to the utilization of grass and other green crops. *British Patent*, 511525.

SMITH A.M. & AGIZA A.H. (1951). The amino acids of several grassland species, cereals and bracken. *J. Sci. Fd Agric.* **2,** 503.

SMITH P., JR, AMBROSE M.E. & KNOBL E.M., JR (1965). Possible interference of fats, carbohydrates, and salts in amino acid determinations in fish meals, fish protein concentrates, and mixed animal feeds. *J. agric. Fd Chem.* **13,** 266.

SMITH R.H. (1966). Lipid-protein isolates. *Adv. Chem. Ser.* No. 57, p. 133. American Chemical Society, Washington.

SMYTH D.G. & ELLIOTT D.F. (1964). Some analytical problems involved in determining the structure of proteins and peptides. *Analyst,* **89,** 81.

SPENCER R.R., BICKOFF E.M., KOHLER G.O., WITT S.C., KNUCKLES B.E. & MOTTOLA A. (1970). Alfalfa products by wet fractionation. *Trans. Am. Soc. agric. Engng,* **13,** 198.

SPIES J.R. (1967). Determination of tryptophan in proteins. *Analyt. Chem.* **39,** 1412.

SPIES J.R. & CHAMBERS D.C. (1948). Chemical determination of tryptophan. Study of color-forming reactions of tryptophan, p-dimethylaminobenzaldehyde, and sodium nitrite in sulfuric acid solution. *Analyt. Chem.* **20,** 30.

SPIES J.R. & CHAMBERS D.C. (1949). Chemical determination of tryptophan in proteins. *Analyt. Chem.* **21,** 1249.

STAHMANN M.A. (1968). The potential for protein production from green plants. *Econ. Bot.* **22,** 73.

STEWARD F.C., WHETMORE R.H., THOMPSON J.F. & NITSCH J.P. (1954). A quantitative chromatographic study of nitrogenous components of shoot apices. *Am. J. Bot.* **41,** 123.

SUBBA RAO M.S., SINGH N. & PRASANAPPA G. (1967). Preservation of wet leaf protein concentrates. *J. Sci. Fd Agric.* **18,** 295.

SUBBA RAU B.H., MAHADEVIAH S. & SINGH N. (1969). Nutritional studies on whole-extract coagulated leaf protein and fractionated chloroplastic and cytoplasmic proteins from lucerne (*Medicago sativa*). *J. Sci. Fd Agric.* **20,** 355.

SUBBA RAU B.H. & SINGH N. (1970). Studies on nutritive value of leaf protein from lucerne (*Medicago sativa*): Part II—Effect of processing conditions. *Indian J. exp. Biol.* **8,** 34.

SWAMINATHAN M.S. (1967). Protein hunger and threat of intellectual dwarfism. *Fd Industries J.* **1,** 4.

TANNENBAUM S.R., BARTH H. & LEROUX J.P. (1969). Loss of methionine in casein during storage with autoxidizing methyl linoleate. *Agric. Fd Chem.* **17,** 1353.

TAYLOR R.D., KOHLER G.O. & MADDY K.H. (1968). Parametric linear programming evaluations of alfalfa meal in poultry and swine rations. *Tenth Tech. Alf. Conf. Proc.* ARS–74–46, 80.

TRACEY M.V. (1948). Leaf protease of tobacco and other plants. *Biochem. J.* **42,** 281.

TRISTRAM G.R. (1966). Preparation of hydrolysates. *Techniques in Amino Acid Analysis,* p. 61, Technicon Instruments Co. Ltd, Chertsey, Surrey.

VALLI DEVI A., RAO N.A.N. & VIJAYARAGHAVAN P.K. (1965). Isolation and composition of leaf protein from certain species of Indian flora. *J. Sci. Fd Agric.* **16,** 116.

VINCENT-CHANDLER J., SILVA S. & FIGARELLA J. (1959). The effect of nitrogen fertilization and frequency of cutting on the yield and composition of three tropical grasses. *J. Agron.* **51,** 202.

WATERLOW J.C. (1962). The absorption and retention of nitrogen from leaf protein by infants recovering from malnutrition. *Br. J. Nutr.* **16,** 531.

WESTLAKE D.F. (1963). Comparisons of plant productivity. *Biol. Rev.* **38,** 385.

WESTLAKE D.F. (1969). Macrophytes. *A Manual on Methods for Measuring Primary Production in Aquatic Environments* (Ed. by Vollenweider), Section 2: 2, p. 25. IBP Handbook No. 12. Blackwell Scientific Publications, Oxford.

W.H.O. (1967). Requirements of vitamin A, thiamin, riboflavin and niacin. *Tech. Rep. Ser.* **362.**

WILSON R.F. & TILLEY J.M.A. (1965). Amino acid composition of lucerne and of lucerne and grass protein preparations. *J. Sci. Fd Agric.* **16,** 173.

WINTERSTEIN E. (1901). Ueber die stickstoffhaltigen Bestandtheile grüner Blatter. *Ber. dtsch. bot. Gess.* **19,** 326.

WOODHAM A.A. (1965). The nutritive value of leaf protein concentrates. *Proc. Nutr. Soc.* **24,** xxiv.

WOODHAM A.A. (1970). Unpublished results.

WOODHAM A.A. & DAWSON R. (1968). The nutritive value of groundnut proteins. I. Some effects of heat upon nutritive value, protein composition and enzyme inhibitory activity. *Br. J. Nutr.* **22,** 589.

YEMM E.W. & FOLKES B.F. (1953). The amino acids of cytoplasmic and chloroplastic proteins of barley. *Biochem. J.* **53,** 700.

YOUNT J.L. & CROSSMEN R.A., JR (1970). Eutrophication control by plant harvesting. *J. Water Poll. Cont. Fed.* **42,** 173.

# List of Participants and Contributors

R.M. Allison, Department of Scientific and Industrial Research, Applied Biochemistry Division, Crop Research Division, Private Bag, Christchurch, New Zealand
D. Bagchi, Indian Statistical Institute, 203 Barrackpore Trunk Road, Calcutta 35, India
R.N. Datta, Ministry of Food and Agriculture, Department of Food, Shastri Bhavan, 35 Haddows Road, Madras 6, India
M.A. Fafunso, Department of Biochemistry, University of Ibadan, Ibadan, Nigeria
K.G. Gollakota, College of Basic Sciences and Humanities, UP Agricultural University, Pantnagar, Nainital Dt. India
K.G. Gunetileke, C.I.S.I.R., P.O. Box 787, Colombo, Ceylon
R.N. Joshi, Department of Botany, Marathwada University, Aurangabad, India
G. Kamalanathan, Sri Avinashilingam Home Science College, Coimbatore 11, India
I. Kenyeres, Budapest XIII, Raul Wallenberg U.2, Hungary
S.G.S. Khanna, College of Basic Sciences and Humanities, UP Agricultural University, Pantnagar, Nainital Dt. India
G.O. Kohler, U.S.D.A., 800 Buchanan Street, Albany, California 94710, U.S.A.
P. Kralovanszky, Protein Bureau of the Hungarian National Development Committee, Budapest XI, Menesi UT Y3, Hungary
S. Matai, Indian Statistical Institute, 203 Barrackpore Trunk Road, Calcutta 35, India
S.R. Mudambi, Sir Vithaldas Thakersey College of Home Science, 1 Nathibai Thakersey Road, Fort, Bombay 20 BR. India
S. Natarajan, Department of Food, Ministry of Food, Agriculture, Community Development and Cooperation, Government of India, New Delhi, India
O.L. Oke, Chemistry Department, University of Ife, Ile-Ife, Nigeria
N.G. Perur, Department of Chemistry and Soils, Agricultural College, Hebbal, Bangalore 24, India
L. Phadnis, College of Home Science, University of Udaipur, Udaipur, India
N.W. Pirie, Biochemistry Department, Rothamsted Experimental Station, Harpenden, Herts., United Kingdom
K. Ramakrishnan, The State Agricultural College and Research Institute, Coimbatore 3, India
K.V.R. Ramana, Central Food Technological Research Institute, Mysore 2(A), India
S.C.Ramdas, The State Agricultural College and Research Institute, Coimbatore 3, India
K.K.P.N. Rao, FAO of the United Nations, 1 Ring Road, Kilokri, New Delhi 14, India
B.H. Subba Rau, Central Food Technological Research Institute, Mysore 2(A), India
P.R. Reddy, Department of Home Science, S.V. University College, Tirupathi, A.P., India

N. Singh, Protein Technology Discipline, Central Food Technological Research Institute, Mysore 2(A), India

J. Tear, Alfa-Laval, AB, Postfack, S-147 00 Tumba, Sweden

M.K. Urs, Central Food Technological Research Institute, Mysore 2 (A), India

A.A. Woodham, Rowett Research Institute, Bucksburn, Aberdeen, AB9 28B, United Kingdom

Those not present at the meeting, who presented papers, were:

D.B. Arkcoll, Biochemistry Department, Rothamsted Experimental Station, Harpenden, Herts., United Kingdom

C.E. Boyd, Savannah River Ecology Laboratory, P.O. Box A, Aitken, South Carolina 29801, U.S.A.

M. Byers, Biochemistry Department, Rothamsted Experimental Station, Harpenden, Herts., United Kingdom

M. Holden, Biochemistry Department, Rothamsted Experimental Station, Harpenden, Herts., United Kingdom

J. Holló, Institute of Agricultural Chemical Technology, University of Technical Sciences, Gellért Tér 4, Budapest XI, Hungary

K. Lexander, Department of Plant Physiology, University of Lund, Lund, Sweden

A. Pirie, Nuffield Laboratory of Ophthalmology, Walton Street, Oxford, United Kingdom

# Conclusions

Preliminary drafts of most of the papers printed here were circulated and discussed at an IBP Technical Group meeting in Coimbatore on 25, 26 and 27 November, 1970. Several other papers were also circulated and discussed and the conclusions from them are incorporated in the final versions of the printed papers. It is this material that is referred to in the text by an author's name followed by (unpublished). At the end of the meeting, four sub-committees drew up recommendations for further work. The whole group discussed these recommendations, but did not have an opportunity to agree this edited version of them.

*Agronomy*

1. (a) The amenability of existing crop plants should be studied.
   (b) Many papers on protein extraction from non-agricultural plants have been published, but thousands of plants remain unstudied. The protein in some of these may extract well and be palatable, they may be amenable to successive cuttings and may be photo-synthetically very efficient. Therefore, this survey work should be continued and extended.
   (c) Species found promising should be grown on replicated small plots with adequate fertilizer, and their regrowth ability explored using standard equipment for assessing yields.
2. Agronomic trials should investigate:
   (a) Spacing.
   (b) Fertilizer treatments.
   (c) Effects of different frequencies of cutting.
   (d) Spraying, micronutrients and chemicals such as simazine.
   (e) Crops should be sown at different seasons because it is possible that growth at an unusual time of year will enhance vegetative growth.
3. Conventional crops which produce by-product leaves should be studied.

**Processing and economics**

1. It is generally accepted that a growing demand for different forms of protein is to be expected.
2. Hitherto leaf protein research has been mainly directed to developing and improving equipment and methodology adapted for small-scale production for human consumption. Work with this equipment should be encouraged and extended.
3. Future effort should also be directed towards industrial scale ventures to produce both fodder and LPC that is nutritionally adequate for human consumption. Towards this end large-scale feasibility studies should be made in various regions.

**Evaluation**

1. A standard biological test for LPC should be adopted, the most suitable is the AOAC method for PER estimation with rats. Both PER and weight should be reported.

   In the absence of reliable data from a standard test we recommend the following tentative levels. These would, however, be modified in the light of future collaborative work.

   PER $>1{\cdot}5$
   Weight gain 25 g in 4 weeks
   Material soluble in water $<1\%$
   Ash $<3\%$ and acid-insoluble ash $<1\%$

2. Standard samples of LPC should be distributed (from a common source) for collaborative work.

*Topics for research*

1. Methods for available lysine estimation suitable for LPC.
2. Studies on carotene content and stability.
3. Shelf-life studies.
4. Factors affecting quality using the standard PER technique.

**Acceptance**

1. A tentative specification of a food grade leaf protein should be laid down in consultation with different workers at the national and international level.

2. Any production unit should meet requirements laid down by the health authorities.
3. The products so produced should be tested in nutrition laboratories, Home Science Colleges and other institutions.
   (a) Food preparations, which are commonly and frequently consumed in the region and which will lend themselves to the incorporation of leaf protein, should be chosen.
   (b) The objective is to ensure the consumption by adults of at least 10 g of leaf protein (i.e. 6 g of 100% protein) per day. Children, according to their weight, would get less. Efforts should be directed towards including this amount in one serving.
   (c) The recipes thus developed should be tested for acceptability by small and large groups.
4. Pilot feeding programmes should be organized to demonstrate the long-term acceptability of leaf protein and the nutritional benefit of using it.
5. Home Science Associations and allied organizations should constitute national working groups to implement these suggestions with the necessary financial assistance from governments and other agencies.

# Index of Plants

# General Index